SHAPE SELECTIVE CATALYSIS IN INDUSTRIAL APPLICATIONS

T0203723

CHEMICAL INDUSTRIES

A Series of Reference Books and Textbooks

Consulting Editor

HEINZ HEINEMANN
Berkeley, California

65. *Shape Selective Catalysis in Industrial Applications: Second Edition, Revised and Expanded*, N. Y. Chen, William E. Garwood, and Francis G. Dwyer

ADDITIONAL VOLUMES IN PREPARATION

Hydrocracking Science and Technology, Julius Scherzer and Adrian Gruia

Catalysis of Organic Reactions, edited by Russell Malz

Hydrotreating Technology for Pollution Control: Catalysts, Catalysis, and Processes, edited by Mario Occelli and Russell Chianelli

Synthesis of Porous Materials: Zeolites, Clays, and Nanostructures, edited by Mario Occelli and Henri Kessler

Methane and Its Derivatives, Sunggyu Lee

SHAPE SELECTIVE CATALYSIS IN INDUSTRIAL APPLICATIONS

Second Edition, Revised and Expanded

N. Y. Chen

Retired from
Mobil Research and Development Corporation
Princeton, New Jersey

William E. Garwood
Francis G. Dwyer

Retired from
Mobil Research and Development Corporation
Paulsboro, New Jersey

CRC Press
Taylor & Francis Group
Boca Raton London New York

CRC Press is an imprint of the
Taylor & Francis Group, an **informa** business

CRC Press
Taylor & Francis Group
6000 Broken Sound Parkway NW, Suite 300
Boca Raton, FL 33487-2742

First issued in paperback 2019

ISBN-13: 978-0-8247-9737-9 (hbk)
ISBN-13: 978-0-367-40129-0 (pbk)

Library of Congress Cataloging-in-Publication Data

Chen, N. Y.
    Shape selective catalysis in industrial applications / N. Y. Chen, William E. Garwood, Frank G. Dwyer.—2nd ed., rev. and expanded.
      p. cm.—(Chemical industries; 65)
    Includes bibliographical references and indexes.
    ISBN 0-8247-9737-X (hardcover: alk. paper)
    1. Catalysis. 2. Zeolites. I. Garwood, William E. II. Dwyer, Francis G. III. Title. IV. Series: Chemical industries; v. 65.
TP156.C35C459   1996
660'.2995—dc20
                      96–15466
                      CIP

Visit the Taylor & Francis Web site at
http://www.taylorandfrancis.com

and the CRC Press Web site at
http://www.crcpress.com

Preface to the Second Edition

Over the past few decades, shape selective catalysis research has grown in many directions. For example, it has expanded from primarily acid catalysis to include acid/metal catalysis and H_2O_2 oxidation over titanium-containing zeolites, and from dealing with strictly small and medium pore zeolites to the incorporation of several large pore zeolites. Discoveries in catalytic process technology from one industry have cross-fertilized new developments in other, seemingly unrelated, industries. We are gratified to see that our work has played a role in this continuously growing field.

More than 2400 articles on medium pore zeolites alone have been listed in the API literature abstracts since 1989. *Derwent World Patent Index* for the same time span lists 1581 patents on medium pore zeolites. A second edition of *Shape Selective Catalysis in Industrial Applications* reflecting the latest developments in the technology and broadening the scope of the first edition was clearly called for. This second edition retains from the first edition introductory material relating the basic principles of shape selective catalysis to their practical applications. New knowledge on the fundamentals of zeolite catalysis and the specific features that make the zeolites shape selective has been added. Reactions carried out using shape selective catalysts have been updated, and known processes using shape selective zeolites have also been updated to include current applications and the future potential of shape selective catalysis in petroleum processing, aromatics processing, alternative fuels, and other new opportunities.

Among the new or expanded topics in this second edition are the following. In Chapter 2, "Relation Between Zeolite Structure and Its Catalytic Activity," we broadened the range of zeolites covered and added many new sections, including ones on 12-membered oxygen ring systems, mesopore systems, and new catalytic characterization tests. Chapter 4 has been broadened and renamed "Shape Selective Catalysis." In addition to acid catalysis we have included acid/metal catalysis and the selective oxidation catalysis with titanium zeolites. Chapter 5, "Applications in Petroleum Processing," has been revised to cover two major topics: (1) the role shape selective catalysis can play in providing "environmentally clean" fuels and (2) the technologies in dewaxing for clean distillate and jet fuels and new lube basestocks. Chapter 6, "Applications in Aromatics Processing," was updated to reflect recent developments, and Chapter 7, "Applications in Alternate Fuels and Light Olefins," contains a new section on the synthesis of ethers. In Chapter 8, "New Opportunities in Shape Selective Catalysis," we added new material including discussions on the development of zeolitic membrane reactors and environmental applications. Also included are recent studies by Bill Garwood and his co-workers on the effect of adding ZSM-5 to the diet of live animals.

It is a credit to Mobil's management over the years that they have taken the risk to encourage zeolite synthesis and the development of some of the most important processes described in this book. Recently, in order to be more profitable, the industry has endeavored to reduce staff support costs, including research, and to focus on research more closely aligned with the business. It is my personal feeling that in this kind of environment, although the industry may be more profitable for the short term, truly new technologies will be hard to generate.

On behalf of my co-authors Bill Garwood and Frank Dwyer, I wish to thank our former colleagues at Mobil who took the time to review and comment on the manuscript and the courtesy of Mobil for the use of their libraries. We also thank Ruth Henderson, Jane Bitter, Marilyn Wittlinger, and Jean Mannix for their assistance.

N. Y. Chen

Preface to the First Edition

In the course of my 29-year career with Mobil, I have been fortunate enough to witness and play a role in the evolution of shape selective catalysis from a relatively straightforward concept into a major branch of industrial catalytic process technology. During this period, many new zeolite-based processes have entered into commercial practice in the petroleum, petrochemical, and chemical industries. The majority of these have been based on the shape selective attributes of these zeolites. With the technology of catalytic shape selectivity having reached a certain degree of maturity, it seemed an appropriate juncture to write about the principles of shape selective catalysis and the industrial application of shape selective zeolites.

The idea for this undertaking was started over ten years ago at the urging of Heinz Heinemann, the founding editor of *Catalysis Reviews* and an old colleague, now at the University of California—Berkeley. Bill Garwood and I, together with Werner Haag and Al Schwartz, wrote two general review papers on the subject of hydrocarbon conversion over shape selective zeolite catalysts. These papers were presented at the First Symposium on Advances in Catalytic Chemistry in Snowbird, Utah, and at the AIChE 72nd Annual Meeting in San Francisco, California, in 1979.

It took Bill and me more than five years to complete the next phase of our project, which was to review the rapidly expanding number of journal articles and patents describing the industrial applications of shape selective catalysis. This work was published in *Catalysis Reviews—Science and Engineering* in 1986. The present work is an updated and greatly expanded version of this review paper. We've had the good fortune of adding Frank Dwyer as a coauthor. Frank, who has over 30 years of experience in the area of zeolite catalysis, has contributed significantly to the content of the first four chapters.

This book is divided into nine chapters. Introductory material, including a brief presentation of the fundamentals of zeolite catalysis, constitutes the first three chapters. We hope that these chapters will be useful to the reader in bridging the gap between the basic principles of shape selective catalysis and their practical applications.

The major portion of this book, Chapters 4–8, is devoted to shape selective reactions catalyzed by zeolites and their industrial applications. Attempts have been made to include in Chapter 4 all the hydrocarbon and non-hydrocarbon reactions catalyzed by shape selective zeolites. Chapters 5 and 6 describe commercialized or developed processes in the petroleum and petrochemical industries, respectively. Chapter 7 describes the new processes for the production of fuels and chemicals from nonconventional feedstocks. Included in this chapter is an updated description of Mobil's MTG process for the production of gasoline from methanol, a process that has already been commercialized in New Zealand. Chapter 8 covers a diverse field of potential opportunities for the application of shape selective catalysis including: oil production, shale oil, coal, natural gas upgrading, internal combustion engine modification, biomass conversion, and applications in the fermentation, chemical, and waste recovery industries.

Judging from the steady increase in the number of zeolite process application patents filed each year in the U.S. Patent Office and patent offices around the world, there is little doubt that the application of shape selective catalysis will continue to grow and extend to metal and bifunctional catalyzed reactions as well as to the commercialization of non-hydrocarbons processes and the synthesis of fine chemicals. Frequently, discoveries from one industry fertilize new developments in other, seemingly unrelated industries. Thus, it is the fervent hope of the authors that this book will serve as a bridge of "technology transfer" among various industries. Most importantly, however, we hope that this book will serve as a useful reference for workers in the field.

The advent of shape selective catalysis can be traced back to the discovery of the unique catalytic selectivity of Zeolite A by Paul Weisz and Vince Frilette at Mobil in the late 1950s. The discovery, although exciting to a few of us, did not set the world on fire. In fact, at the time there was not a zeolite that could withstand the rigors of the industrial environment and be considered of practical value. A small group of zeolite enthusiasts labored for five years in relative obscurity before making a practical breakthrough, which led to the commercialization of the first catalytic process, the Selectoforming process, using a stable shape selective zeolite as the catalyst.

We at Mobil owe a debt of gratitide to my mentor, Paul Weisz, for his conviction and his foresight in emphasizing the significance of shape selective catalysis. He reminded us of the similarity of shape selective catalysis to enzyme catalysis, where the geometric shape of a molecule and its environment play important roles. Admittedly, the activity and specificity of shape selective zeolite catalysts are still orders of magnitude less than those of enzyme catalysts.

It is also the credit of Mobil's management to have taken the risk, marshalling at times the entire staff of two laboratories, for the development of some of these processes described in this book. The announcement of the MTG process in 1976 caught the imagination of the academic world and stimulated fundamental research on medium pore zeolites and C_1 chemistry. This is evidenced by the explosive proliferation of technical papers on these subjects, which, in turn, has greatly expanded our knowledge base. Hopefully, a careful study of where shape selective zeolite catalysis has been applied will lead to additional industrial applications.

On behalf of my coauthors, Bill Garwood and Frank Dwyer, I wish to thank many of our colleagues who took the time to review and comment on the manuscript. Mrs. Delma Dow, Mrs. Diane Tavin, and Mrs. Marianne Snyder deserve special recognition for undertaking the administrative details associated with preparing the manuscript.

N. Y. Chen

Contents

SHAPE SELECTIVE CATALYSIS IN INDUSTRIAL APPLICATIONS

1

Introduction

Over thirty-five years ago, Weisz and Frilette (1960) coined the term *shape selective catalysis* to describe the then unexpected intrinsic catalytic activities of synthetic crystalline molecular sieve zeolites. They found that the calcium ion exchanged zeolite A, having "port" sizes of 4 to 5 Å selectively cracked straight chain *n*-paraffins to exclusively straight chain products. Since that time, great strides in the use of shape selective catalysts in commercial catalytic processes have been made. Their applications have expanded well beyond the boundaries of traditional petroleum refining to petrochemical and chemical manufacturing.

Over the past 35 years, many new synthetic zeolites have been discovered. Of particular importance was the discovery of the synthetic medium pore zeolites, having "port" sizes of 5 to 6 Å. These include, among others, ZSM-5 (Argauer and Landolt, 1972), ZSM-11 (Chu, 1973), ZSM-23 (Plank et al., 1978; Valyocsik, 1984b), ZSM-35 (Plank et al., 1977), ZSM-48 (Chu, 1983; Rollmann and Valyocsik, 1983), NU-6 (Whittam, 1983), and Theta-1 (Ball et al., 1985).*

---

*Not listed are the many zeolite compositions that have been given different nomenclature but are topologically isostructural and otherwise the same as some of the above zeolites, for example, Silicalite, AMS-1, AS-1, CZM, FZ-1, NU-5, TSZs, TZs, ZBMs, ZMQs, and ZSM-5, ISI-1, KZ-2, NU-10, ZSM-22, and Theta-1.

While classical zeolites contain Si and Al, the addition of the elements Ga, Ge, Be, B, Fe, Cr, P, and Mg in framework positions has also been achieved. Indeed, the number of elements that can be inserted into the crystal framework either by isomorphous substitution or by direct synthesis has grown dramatically in recent years (Flanigan et al., 1986).

The availability of synthetic zeolites has greatly expanded the realm of "shape selectivity." Since the discovery of selective conversion of straight chain molecules, it has become possible to selectively convert such molecules as certain branched molecules, single ring aromatics, naphthenes, and nonhydrocarbons with a critical molecular dimension less than about 6 Å.

During this period, major technological advances have also resulted from research and development efforts at Mobil and other research laboratories. The commercial production of ZSM-5, the commercialization of new petroleum refining and petrochemical processes, and the ZSM-5 catalyzed methanol-to-gasoline (MTG) process have attracted worldwide attention. Interest in medium pore zeolites has grown in recent years, both at academic institutions and at other industrial laboratories. This is evidenced by the exponential growth of publications dealing with the fundamental and applied aspects of shape selective catalysis. Molecular shape selective zeolites have gone from being laboratory curiosities to becoming commercially significant groups of industrial catalysts.

More recently shape selective catalysis has extended from the medium pore zeolites to several large pore zeolites, including ZSM-4 (Chen, 1986), zeolite L (Derouane and Vanderveken, 1988; Martens et al., 1988) and zeolite Beta (LaPierre et al., 1983; Weitkamp et al., 1989).

Through the isomorphous substitution of silicon by titanium in ZSM-5, a titanium-containing zeolite, TS-1, a novel derivative of ZSM-5, was first synthesized by the researchers at ENI of Italy (Taramasso et al., 1983). A remarkable class of reactions having industrial importance, the H_2O_2 oxidation of alkenes and aromatics in the production of hydroquinone and catechol, has been developed (Notari, 1987).

The ability to produce fuels and chemicals from unconventional raw materials by shape selective catalysts is exemplified by obtaining light olefins and gasoline from oxygen-containing feedstock such as methanol, which can be produced from these unconventional raw materials.

Similar applications in the fine chemicals and pharmaceuticals industries can also be expected. Curiously, development in these two areas has remained fairly dormant, at least to the outside world. Although the petroleum industry is primarily concerned with hydrocarbon fuels and lubes, and not with the fine chemicals, technology transfer between these two industries should benefit each. There is the need to transfer technology efficiently and form interindustry technical alliances to promote the research and development of new technology—not only to provide improved catalysts for existing process technology, but also to create totally new processes and new products to improve our national competitiveness.

Over the years, the principles of shape selective catalysis have been comprehensively reviewed by a number of authors (Venuto and Landis, 1968; Weisz, 1973; Csicsery, 1976; Chen et al., 1979a,b; Weisz, 1980; Derouane, 1980; Heinemann, 1981; Derouane, 1982; Whyte and Dalla Betta, 1982; Dwyer and Dyer, 1984; Haag, 1984a,b; Csicsery, 1984, 1986; Hölderich et al., 1988; Hölderich, 1991; Venuto, 1994; Dartt and Davis, 1995). Our purpose here is not to review the subject matter in general, but to summarize the impact of shape selective catalysis on the petroleum and petrochemical industries.

To provide a brief introduction to the fundamentals of zeolite catalysis, we have included two short chapters that describe the relation between catalyst structure and catalytic activity and the principal methods of achieving molecular shape selectivity. We hope these chapters will be useful to the reader in bridging the gap between the basic principles of shape selective catalysis and their practical applications. The main emphasis of this review can be found in Chapters 4 through 8.

REFERENCES

Argauer, R. J., and G. R. Landolt, U.S. Pat. 3,702,886, Nov. 14, 1972.

Ball, W. J., S. A. I. Barri, and D. Young, U.S. Pat. 4,533,649, Aug. 6, 1985 (UK Appl 8,225,524, Sep. 7, 1982).

Chen, N. Y., Proc. 7th Int. Zeol. Conf., Y. Murakami, A. Iijima, and J. W. Ward, eds., Kodadansha/Elsevier, Tokyo/Amsterdam, p. 653 (1986).

Chen, N. Y., W. E. Garwood, W. O. Haag, and A. B. Schwartz, "Shape Selective Hydrocarbon Catalysis Over Synthetic Zeolite ZSM-5," paper presented at Symp. Advan. Catal. Chem. I, Snowbird, Utah, Oct. 3, 1979a.

Chen, N. Y., W. E. Garwood, W. O. Haag, and A. B. Schwartz, "Shape Selective Conversion Over Intermediate Pore Size Zeolite Catalysis," paper presented at the AIChE 72nd Annual Mtg., San Francisco, Nov. 25–29, 1979b.

Chu, P., U.S. Pat. 3,709,979, Jan. 9, 1973.

Chu, P., U.S. Pat. 4,397,827, Aug. 9, 1983.

Csicsery, S. M., *Zeolite Chemistry and Catalysis*, J. A. Rabo, ed., Am. Chem. Soc. Monogr. *171*, 680 (1976).

Csicsery, S. M., Zeolites *4*, 202 (1984).

Csicsery, S. M., Pure Appl. Chem. *58*, 841 (1986).

Dartt, C. B. and M. E. Davis, Catal. Today *19*, 151 (1995).

Derouane, E. G., Stud. Surf. Sci. Catal. *5*, 5 (1980).

Derouane, E. G., *Intercalation Chemistry*, W. M. Stanley and A. J. Jacobson, eds., Academic Press, New York, p. 101 (1982).

Derouane, E. G., D. Vanderveken, Appl. Catal. *45*, L15 (1988).

Dwyer, J., and A. Dyer, Chem. Ind. *265*, 237 (1984).

Flanigan, E. M., B. M. Lok, R. L. Patton, and S. T. Wilson, Proc. 7th Int. Zeol. Conf., Y. Murakami, A. Iijima, and J. W. Ward, eds., Kodansha/Elsevier, Tokyo/ Amsterdam, p. 103 (1986).

Haag. W. O., *Heterogeneous Catalysis*, Vol. 2, B. L. Shapiro, ed., Texas A & M Univ. Press, College Station, TX, p. 95 (1984a).

Haag. W. O., Proceedings 6th Int. Zeol. Conf., D. Olson and A. Bisio, eds., Butterworths, Surrey, UK, p. 466 (1984b).

Heinemann, H., Catal. Rev.-Sci. Eng. *23*, 315 (1981).

Hölderich, W., M. Hesse and F. Näumann, Agew. Chem. Int. Ed. Engl. *27*, 226 (1988).

Hölderich, W. F., Stud. Surf. Sci. Catal. *58*, 631 (1991).

LaPierre, R. B., R. D. Partridge, N. Y. Chen, and S. S. Wong, U.S. Pat. 4,419,220, Dec. 6, 1983.

Martens, J. A., J. Perez-Pariente, A. Corma, and P. A. Jacobs, Appl. Catal. *45*, 85 (1988).

Notari, B., Stud. Surf. Sci. Catal. *37*, 413 (1987).

Plank, C. J., E. J. Rosinski, and M. K. Rubin, U.S. Pat. 4,016,245, Apr. 5, 1977.

Plank, C. J., E. J. Rosinski, and M. K. Rubin, U.S. Pat. 4,076,842, Feb. 28, 1978.

Rollmann, L. D. and E. W. Valyocsik, U.S. Pat. 4,423,021, Dec. 27, 1983.

Taramasso, M., G. Perego, and B. Notari, U.S. Pat. 4,410,501, Oct. 18, 1983.

Valyocsik, E. W., U.S. Pat. 4,481,177, Nov. 6, 1984a.

Valyocsik, E. W., U.S. Pat. 4,490,342, Dec. 25, 1984b.

Venuto, P. B., Microporous Mat. *2*, 297 (1994).

Venuto, P. B. and P. S. Landis, Advan. Catal. *18*, 259 (1968).

Weisz, P. B., Chemtech *3*, 498 (1973).

Weisz, P. B., Pure appl. Chem. *52*, 2091 (1980).

Weisz, P. B. and V. J. Frilette, J. Phys. Chem. *64*, 382 (1960).

Weitkamp, J., S. Ernst., C. Y. Chen, Stud. Sci. Surf. Catal. *49*, 1115, 1989.

Whittam, T. V., U.S. Pat. 4,397,825, Aug. 9, 1983 (UK Appl. 8,039,685, Dec. 11, 1980).

Whyte, T. E. and R. A. Dalla Betta, Catal. Rev.-Sci. Eng. *24*, 567 (1982).

2

Relation Between Zeolite Structure and Its Catalytic Activity

I. ZEOLITES

For decades, zeolites have been described as crystalline aluminosilicate molecular sieves that have open porous structures and ion exchange capacities. We are aware now that these materials may contain elements in addition to silicon and aluminum in their framework structures.

Nature has provided us with 34 different zeolites (Barrer, 1968; Breck, 1974; Meier, 1979; Meier and Olson, 1992). But among those of interest to catalysis, only a few are found in abundance and even fewer have found industrial use. The industrial application of zeolite catalysts depends largely on our ability to synthesize zeolites, and the synthesis of known and new structures has made new discoveries in zeolite catalysis possible. Today, not including the more recently discovered "mesopore systems," more than 80 different aluminosilicate zeolite structures are available (von Ballmoos and Higgins, 1990), with pore openings that allow the passage of molecules ranging in size from less than 5 Å to larger than 10 Å. Many crystalline metallophosphate zeolites, either having the same framework structures as aluminosilicates or new structures, have been synthesized (Wilson et al., 1982a,b; Lok et al., 1984, 1985; Wilson and Flanigan, 1986).

The catalytic sites in aluminosilicate zeolites are associated with tetrahedral aluminum atoms in substitutional positions in the framework of silica. In the case of hydrogen zeolites, protons associated with the negatively charged framework aluminum are the source of Brønsted acid activity.

Metallophosphate zeolites, on the other hand, derive their acid activity by the isomorphous substitution in the aluminophosphate framework structure. With equimolar concentration of the trivalent aluminum and pentavalent phosphorus, the parent structure, like the tetravalent silicon in silica, has no intrinsic acidity. An imbalance of aluminum and phosphorus through the replacement of either aluminum or phosphorus by other atoms, such as silicon, results in a charged framework and a source of acidity or basicity.

II. PORE/CHANNEL SYSTEMS

Zeolites of interest to shape selective catalysis may be divided into five major groups according to their pore/channel systems. Some of their structural properties are presented in Table 2.1.

A. 8-Membered Oxygen Ring Systems

These include most of the earliest known shape selective small pore zeolites, such as Linde Type A, erionite, and chabazite. Other members of this group include ZK-5, ZSM-34 (Rubin et al., 1978), and high silica analogs of Linde Type A, that is, zeolite Alpha; ZK-4, ZK-21, and ZK-22, and several other less common natural zeolites.

Among the many metallophosphate zeolites ($AlPO_4$, SAPO, TAPO, MeAPO, etc.), some have the same structures as zeolites; for example, SAPO-17 has the same structure as erionite, SAPO-34 same as chabazite, SAPO-35 same as levynite, SAPO-42 same as zeolite A, and SAPO-43 same as gismondine. While others such as SAPO-14, 18, 26, 33, 44, and 47 are believed to be small pore materials. The shape of the 8-membered oxygen rings varies from circular to puckered and elliptical. The dimension of the pore opening also varies accordingly. For example, Linde Type A and ZK-4 have circular openings, while erionite and chabazite have puckered and elliptical pore openings.

Members of this group sorb straight chain molecules such as *n*-paraffins and olefins and primary alcohols. It is noted that these molecules have critical dimensions larger than the pore size values derived from crystallographic data. Table 2.2 shows an example of such a comparison. The reason for such an apparent incongruity is that these dimensions are calculated on the basis of hard spheres, and in reality the effective pore size depends on the interaction of the

Table 2.1 Pore Structure of Zeolites

IUPAC code	Pore system	Number of rings	Pore size, Å	Description
I. 8-Membered pore system				
CHA	Chabazite	8	3.8 × 3.8	intersecting
RHO	RHO	8	3.6	intersecting
KFI	ZK-5	8	3.9	intersecting
ERI	Erionite	8	3.6 × 5.1	intersecting
	ZSM-34[1]	8	?	intersecting
LTA	Linde Type A (ZK-4)	8	4.1	intersecting
II. Medium pore system				
PAR	Partheite	10	3.5 × 6.9	one-dimensional
LAU	Laumontite	10	4.0 × 5.3	one-dimensional
AEL	SAPO-11	10	3.9 × 6.3	one-dimensional
TON	Theta-1 (ZSM-22)	10	4.4 × 5.5	one-dimensional
MTT	ZSM-23	10	4.5 × 5.2	one-dimensional
	ZSM-48[2]	10	5.3 × 5.6	one-dimensional
EUO	ZSM-50 (EU-1)	10	4.1 × 5.7	one-dimensional
NES	NU-87	10	4.7 × 6.0	2D intersecting
MFI	ZSM-5	10	5.3 × 5.6, 5.1 × 5.5	intersecting
MEL	ZSM-11	10	5.3 × 5.4	intersecting
III. Dual pore system				
FER	Ferrierite (ZSM-35. FU-9)	10	4.2 × 5.4	one-dimensional
		8	3.5 × 4.8	10:8 intersecting
HEU	Heulandite/ clinoptilolite	10, 8	3.0 × 7.6, 3.3 × 4.6	one-dimensional
		8	2.6 × 4.7	10:8 intersecting
MFS	ZSM-57	10	5.1 × 5.4	one-dimensional
		8	3.3 × 4.8	10:8 intersecting
STI	Stilbite	10	4.9 × 6.1	one-dimensional
		8	2.7 × 5.6	10:8 intersecting
OFF	Offretite	12	6.7	one-dimensional
		8	3.6 × 4.9	12:8 intersecting
MOR	Mordenite	12	6.5 × 7.0	one-dimensional
		8	2.6 × 5.7	12:8 intersecting
GME	Gmelinite	12	7.0	one-dimensional
		8	3.6 × 3.9	two-dimensional
MAZ	Mazzite (ZSM-4)	12, 8	7.4, 3.4 × 5.6	one-dimensional
	MCM-22[3]	12	7.1	capped by 6 rings
		10	elliptical	two-dimensional

Table 2.1 *Continued.*

IUPAC code	Pore system	Number of rings	Pore size, Å	Description
BOG	Boggsite	12, 10	7, 5.2 × 5.8	intersecting
	SSZ-26[4]	12, 10		
	SSZ-35[5]	12, 10		

IV. Large pore system

MTW	ZSM-12	12	5.5 × 5.9	one-dimensional
BEA	Beta	12	7.6 × 6.4, 5.5 × 5.5	intersecting
LTL	Linde Type L	12	7.1	one-dimensional
EMT	EMC-2	12	7.1	one-dimensional
		12	7.4 × 6.5	12:12 intersecting
FAU	Faujasite	12	7.4	intersecting
		12	7.4 × 6.5	12:12 intersecting

V. Meso pore system

| VFI | VPI-5 | 18 | 12.1 | one-dimensional |
| | M41-S | | 16 to 100 | one-dimensional |

[1]A member of offretite-erionite family (Rubin et al., 1978).
[2]A disordered structure consisting of ferrierite sheets linked via bridging oxygen located on mirror planes (Schlenker et al., 1985).
[3]A new boro- and aluminosilicate zeolite with two-dimensional sinusoidal 10-ring channels and a 12-ring channel (Leonowicz et al., 1994).
[4]A new aluminosilicate zeolite with intersecting 10- and 12-ring pores, intermediate in size to ZSM-5 and beta (Lobo et al., 1994).
[5]A borosilicate similar in structure to SSZ-33 (Lobo et al., 1994).
Source: Meier and Olson (1992).

Table 2.2 Comparison of Pore Size and Critical Dimension of Sorbate Crystallographic Pore Size (Å)

Linde 5A	4.1
Erionite	3.6 × 5.2
Chabazite	3.6 × 3.7

Molecular dimensions[a]

n-Hexane	3.9 × 4.3 × 9.1

[a]Estimated from Courtald space filling models.
Source: Data taken from Meier and Olson (1992).

intramolecular and the interatomic forces between the host structure and the diffusing molecules. Thus the actual pore size can be significantly larger than that calculated (Wu et al., 1986).

The pore/channel systems of these zeolites also contain interconnecting "supercages," which are much larger than the size of the connecting "windows." The supercage/window structure is often blamed for the cause of catalyst deactivation or coking of acidic catalysts. Bulky molecules such as polycyclic aromatics, which are formed in the supercages, cannot escape through the windows. These molecules are trapped within, and end up as coke deposits (Venuto et al., 1966; Venuto and Landis, 1968; Walsh and Rollmann, 1977, 1979; Rollmann and Walsh, 1979, 1982).

The first zeolite used in a commercial molecular shape selective catalytic process, the Selectoforming process (see Chapter 5), was one of these 8-membered oxygen ring opening small pore zeolites, erionite. Figure 2.1 shows the framework structure of erionite (Staples and Gard, 1959; Bennett and Gard, 1967). The zeolite has both large 12-membered oxygen rings and small 8-membered oxygen rings in its framework structure. The large 12-membered oxygen ring channels are formed by bridging columns of cancrinite cages with hexagonal rings. However, these channels are blocked in their longitudinal directions. The only access to these channels are 8-membered oxygen ring openings which are perpendicular to the channels. Therefore, erionite belongs to the 8-membered oxygen ring system, and it sorbs only straight chain molecules.

B. 10-Membered Oxygen Ring Systems

These are also known as medium pore zeolites. Among the varieties of unique crystal structure types in this group are ZSM-5 (Kokotailo et al., 1978a) and ZSM-11 (Kokotailo et al., 1978b), which are known as pentasils,* Theta-1 (Barri et al., 1984), which is isostructural with ZSM-22 (Kokotailo et al., 1985), ZSM-23 (Rohrman et al., 1985), ZSM-48 (Schlenker et al., 1985), NU-87 (Shannon et al., 1991), partheite, and laumontite.

Except for partheite and laumontite, which have puckered 10-membered oxygen rings in their structures, almost all medium pore zeolites of interest to shape selective catalysis are synthetic in origin. Their framework structures contain 5-membered oxygen rings and they are more siliceous than previously known zeolites. In many instances, these zeolites may be synthesized with a predominance of silicon and with only a very small concentration of aluminum

---

*The term *pentasil* was defined by Kokotailo and Meier (1980) as a family of zeolites having similar structures with ZSM-5 and ZSM-11 as its two end members. These framework structures are all formed by linking chains of 5-membered ring secondary building units.

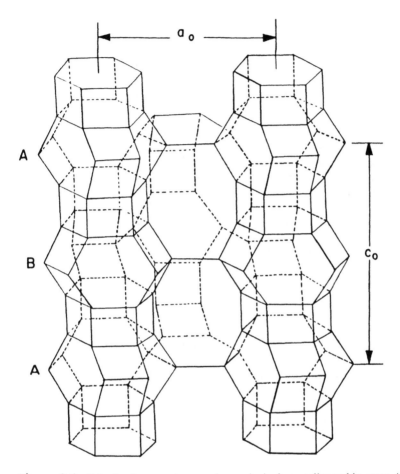

Figure 2.1 Erionite framework. a_o and c_o, principal crystallographic axes; ABA, stacking order of cancrinite cages. (From Gorring, 1973.)

and other atoms. Thus, these zeolites may be considered as "silicates" with framework substitution by small quantities of aluminum and other elements (Dwyer and Jenkins, 1976).

Among the metallophosphate zeolites (AlPO$_4$, SAPO, TAPO, MeAPO, etc.) 11, 30, and 41 are believed to have nonintersecting one-dimensional channel medium-pore structures.

As in the case of the 8-membered oxygen ring zeolites, the shape and size of the 10- or 12-membered oxygen rings also varies from one structural type to

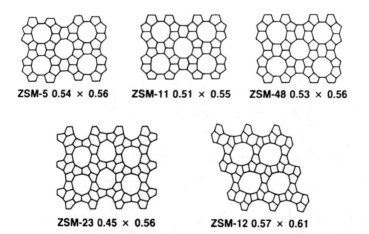

ZSM-5 0.54 × 0.56   **ZSM-11 0.51 × 0.55**   **ZSM-48 0.53 × 0.56**

ZSM-23 0.45 × 0.56   **ZSM-12 0.57 × 0.61**

Figure 2.2   Projections of ZSM-5, -11, -12, -23, and -48 structures. (From Wu et al., 1986.) According to Meier and Olson (1992): ZSM-5: 5.3 × 5.6↔5.1 × 5.5 Å, ZSM-11: 5.3 × 5.4, ZSM-23: 4.5 × 5.2, ZSM-12: 5.5 × 5.9.

another. They range from nearly circular to elliptical to odd shapes such as tear drops (Wu et al., 1986). Figure 2.2 shows the projections of the main channels for some of the zeolites.

Among the zeolites in this group, only ZSM-5, ZSM-11, and NU-87 have intersecting channels. The others have nonintersecting, one-dimensional channels.

ZSM-5 has received the most attention by far and serves as a prime example of the uniqueness of medium pore zeolites in catalysis. ZSM-5 can be synthesized by including organic molecules such as tetrapropylammonium bromide as a directing agent in the reaction mixtures. The organic molecules are incorporated into the zeolite crystals filling the intracrystalline void space either as organic cations or as occluded salt molecules (Dessau et al., 1987) when the zeolite is crystallized from solution (Argauer and Landolt, 1972; Rollmann, 1979a; Kerr, 1981). Other organic reagents which have been used as directing agents in synthesis include amines, diamines, and alcohols.

Within a much narrower compositional space, ZSM-5 can also be synthesized in the absence of an organic directing agent (Plank et al., 1979; Grose and Flanigan, 1981; Wang et al., 1981; Bezak and Mostowicz, 1985; Nastro et al., 1985).

HZSM-5 and HZSM-11 are remarkably stable as acidic catalysts. Unlike other zeolites, they have pores of uniform dimension and have no large supercages with smaller size windows. The absence of bottlenecks in their pore system is believed to be a significant factor for their unusually low coke forming propensity as acidic catalysts. Other contributing factors to their low coking tendencies include their high silica-to-alumina ratios and the geometrical constraint imposed by the 10-membered oxygen ring sized pores.

These factors make it sterically difficult to form the large polynuclear hydrocarbons responsible for coking and irreversible deactivation (Walsh and Rollmann, 1979; Rollmann and Walsh, 1979, 1982; Derouane, 1985). This nonaging feature is probably one of the major reasons for the successful industrial application of these zeolites.

With the aid of a hydrogenation function, some of the one-dimensional, 10-membered oxygen ring zeolites, such as paraffin hydroisomerization over ZSM-22 and ZSM-23 (Chen et al., 1989) and over SAPO-11 (Miller, 1989) are finding industrial applications.

C. Dual Pore Systems

Zeolites in this group have interconnecting channels of 10- and 8-membered oxygen ring openings, or 12- and 8-membered oxygen ring openings, or 12- and 10-membered oxygen ring openings. Examples are ferrierite (Winquist, 1976), ZSM-35 (Plank et al., 1977), FU-9 (Seddon and Whittam, 1985), heulandite/clinoptilolite, ZSM-57 (Schlenker et al., 1990), stilbite, offretite, mordenite, gmelinite, mazzite (ZSM-4, omega), etc. Four zeolites having channels of 12- and 10-membered oxygen ring openings. Boggsite, a rare mineral, was first reported by Pluth and Smith (1990); MCM-22 was reported in 1990 (Rubin and Chu); and the synthetic SSZ-26 and SSZ-33 were reported by Lobo et al. in 1993. More physicochemical characterization studies on these zeolites were published recently (Leonowicz et al., 1994; Lobo et al., 1994).

Some of these dual pore zeolites show interesting catalytic properties in cracking reactions. For example, cracking of paraffins over offretite (Chen et al., 1984) was found to yield more low molecular weight cracked products than that produced from medium pore zeolites. It was speculated that this is because of the availability of the smaller channels which are accessible only by the smaller molecules. However, most of the natural varieties and some of the synthetic samples contain numerous stacking faults, and in many catalytic reactions they behave like small pore zeolites (Miale et al., 1966; Kibby et al., 1974; Chen et al., 1978a).

The potential practical interest in the more recently discovered 12- and 10-membered oxygen ring dual pore systems is showing up in the patent literature, for example, MCM-22 is used in the disproportionation and transalkylation of C_9^+ aromatics and toluene (Absil et al., 1991) and the aromatization of C_2 to C_{12} aliphatics (Chu et al., 1991).

D. 12-Membered Oxygen Ring Systems

Zeolites with 12-membered oxygen ring openings, such as faujasites, are generally known as large-pore zeolites. However, a number of 12-membered oxygen ring zeolites are structurally different and have openings smaller than those of faujasites; these include ZSM-12 (LaPierre et al., 1985; Fyfe et al., 1990), zeolite beta (Higgins et al., 1988), Linde Type L (Barrer and Villiger, 1969), and EMC-2 (Baerlocher, 1990). Applications also appear in the patent literature such as paraffin hydroisomerization and hydrocracking over zeolite beta (LaPierre et al., 1983), catalytic cracking with a mixture of faujasite and zeolite beta (Chen et al., 1990), the liquid phase alkylation over zeolite beta (Innes et al., 1992), the isomerization of alkylaromatics over ZSM-12 (Sachtler et al., 1992), and the high selectivity for aromatization of *n*-hexane over Pt/BaKL (Derouane and Vanderveken, 1988).

E. Mesopore Systems

It was long believed that the upper limit of zeolite pores was 12-membered oxygen rings, although there was no scientific evidence in support of this position. Then Davis et al. (1989) reported the synthesis of VPI-5, an 18-membered oxygen ring structure. The major contribution to the mesopore systems was the work of Kresge et al. (1992) and Beck et al. (1992) when they reported the discovery of the M41-S series of structures whose pore sizes range from 16 to 100+ Å. These structures have uniform unidirectional pores and are synthesized via a liquid crystal templating procedure whereby the pore size can be controlled by the choice of the organic template used in the crystallization mix. Although the original structures were silica only, subsequent synthesis procedures have incorporated alumina and other metals into the structure. In addition, Vartuli et al. (1995) reported that by variations in the syntheses, these mesopore materials can be formed in hexagonal, cubic, or lamellar crystal structures.

III. STRUCTURAL FEATURES OF ZEOLITES AND THE QUANTITATION OF ACIDIC FRAMEWORK ALUMINUM SITES

As described earlier, the medium pore zeolites, unlike other zeolites, have pores of uniform dimension and have no large supercages with smaller size windows. The 10-membered oxygen ring opening channels in medium pore zeolites are intermediate in size as compared with the smaller 8-membered oxygen ring openings and the larger 12-membered oxygen ring openings present in such zeolites as faujasites (zeolite X, zeolite Y, etc.) and mordenite. Materials in the mesopore systems offer potential in the shape selective catalysis of large hydrocarbon molecules as well as in nanochemistry (Ozin, 1992), although no commercial application has yet been developed.

The structure of ZSM-5 was reported by Mobil workers in 1978 (Kokotailo et al., 1978a). The zeolite has been synthesized over a range of silica-to-alumina ratios (Argauer and Landolt, 1972; Dwyer and Jenkins, 1976). Flanigan and co-workers (1978a,b) gave the name "silicalite" to a ZSM-5 having a high silica-to-alumina mole ratio, even though it has the same framework structure.

Highly siliceous zeolites may have ion exchange capacities in excess of their framework tetrahedral aluminum content (Fegan and Lowe, 1984). These aluminum-independent ion exchange sites have been identified as framework siloxy anions that serve as counter-ions for the organic cations but do not contribute to acid activity (Chester et al., 1985).

Of significance to shape selective catalysis is the presence of two intersecting channels formed by rings of 10 oxygen atoms. These two intersecting channels, both formed by 10-membered oxygen rings, are slightly different in their pore size (Meier and Olson, 1992). One sinusoidal channel, which has an elliptical opening (5.1 × 5.5 Å), has the 10-ring parallel to [100]. The other channel, which is straight, has the 10-ring parallel to [010], with a nearly circular (5.3 × 5.6 Å) opening. These two types of channels intersect to form a three-dimensional network of pores.

ZSM-11, another medium pore zeolite has two straight, elliptical channels intersecting at right angles to each other. Both channels are 5.3 × 5.4 Å in pore size (Kokotailo, 1978b). NU-87, a new zeolite found in 1991, has 10-ring channels (4.7 × 6.0 Å) parallel to [100] with two-dimensional intersections.

The other known medium pore zeolites, SAPO-11, Theta-1 (ZSM-22), ZSM-23, and ZSM-48, have only one-dimensional channels. It is interesting to note that although the size and shape of their pores are slightly different, the pores in Theta-1 (ZSM-22) (pore size 4.4 × 5.5 Å) and ZSM-48 (pore size 5.3 × 5.6 Å) are formed by the same set of 5- and 6-membered oxygen rings in the same order of arrangement as that forming the straight channels of ZSM-5 (pore size 5.3 × 5.6 Å), ZSM-11 (pore size 5.3 × 5.4 Å), and ferrierite. The order of arrangement is slightly different for the ZSM-23 pores, which are teardrop shaped (pore size 4.5 × 5.2 Å) (Rohrman et al., 1985).

Among the 12-membered oxygen ring systems, faujasites remain the ones with the largest pore openings. As the data in Table 2.1 shows, ZSM-12, zeolite beta, Linde Type L, and EMC-2 have smaller channel openings. Furthermore, ZSM-12 and Linde Type L have one-dimensional channels.

Over the years, a great deal of research has been devoted to elucidating the nature and the quantitation of acid sites in zeolites (see, for example, Beagley et al., 1984). In addition to the use of X-ray and electron diffraction methods to determine the framework structures of these zeolites, many other complementary techniques have been applied to characterize their pore/channel systems. A variety of physicochemical characterization methods, including FT-IR spectroscopy, MAS-NMR spectroscopy, and thermogravimetric techniques such as temperature programmed desorption (TPD), in addition to acid-base titration, have been developed. For example, temperature programmed desorption analysis was employed by Mikovsky and co-workers to show that NH_4Y samples of different SiO_2/Al_2O_3 ratios contain acid sites of four different strengths, which appear to increase as the number of aluminum atoms neighboring the protonic site decreases (Mikovsky and Marshall, 1976, 1977; Mikovsky et al., 1979). Figure 2.3 shows a typical temperature programmed ammonia desorption (TPAD) curve of an NH_4Y, where the symbols n_0, n_1, n_2, and n_3 refer to the number of nearest neighboring aluminum atoms associated with the acid site.

The value of TPAD as an aid to catalyst characterization, especially as to the number of Brønsted sites, is unquestioned. One significant characteristic of the rate curves for TPAD from NH_4ZSM-5 is the temperature of maximum rate of desorption (T_m). The T_m value is often accepted as a qualitative indication of the strength of the protonic site created by ammonia desorption from the NH_4^+ site. The higher the T_m, the greater the acid strength of the protonic site.

However, it has been recognized for some time that the experimental spectra can be varied (i.e., the number of peaks, their areas, and T_m by varying the operational variables (see, for example, Yushchenko et al., 1986). Furthermore, while it is generally accepted that for high silica zeolites such as NH_4ZSM-5 (Haag et al., 1984) each framework aluminum is responsible for the formation of one protonic site (or the mole ratio of NH_3 to framework aluminum should be 1), NH_3-TPD data on NH_4ZSM-5 reported in the literature continue to show values higher or lower than expected (see, for example, Dima and Rees, 1990; Sawa et al., 1991).

Kapustin et al. (1988) studied the effect of changing the $SiO_2Al_2O_3$ ratio of NH_4ZSM-5 from 25.6 to 280 at a constant sample weight of 300 mg. Their observed T_m values varied from 370 to 442°C. The basic reason for the change of the T_m is the chromatographic adsorption-desorption effects on T_m. Desorbed ammonia molecules are subject to a "chromatographic" effect of further sorption-desorption on the protonic sites.

MAS-NMR spectroscopy has become a powerful tool for the identification of the nature of silicon, aluminum, and other atoms in zeolites. Interestingly, ^{29}Si-NMR studies on zeolite X and Y samples (Engelhardt et al., 1981; Ramdas et al., 1981; Klinowski et al., 1982) show five types of silicon sites cor-

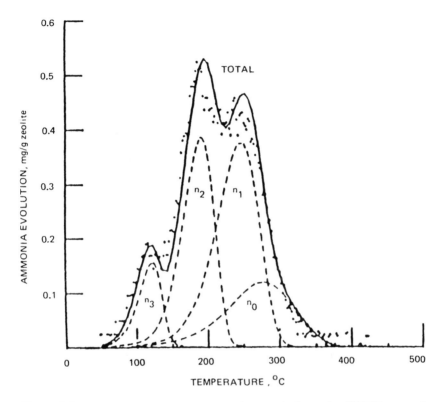

Figure 2.3 Typical temperature-programmed ammonia desorption (TPAD) curve of an NH₄Y. (From Mikovsky and Marshall, 1976.)

responding to the number of nearest aluminum neighbors ranging from 0 to 4. Their population distribution changes with the SiO_2/Al_2O_3 ratio in a manner consistent with the result obtained by TPD analysis. Figure 2.4 shows some typical ${}^{29}Si$-NMR spectra on zeolite NaX and NaY (Engelhardt et al., 1981). Use of probe molecules, such as trimethyl phosphine in combination with MAS-NMR, to identify the nature of acid sites in zeolite Y was reported by Lunsford (1986). In addition to identifying four types of Brønsted acid sites, he found evidence for the development of a multiplicity of Lewis acid sites upon dehydroxylating a HY zeolite.

Sorption measurements using probe molecules of different critical dimensions have been used to gauge the effective size of these pores (Chen and Garwood, 1978; Olson et al., 1981). By the choice of sorbate molecules which have critical dimensions of different shapes (for example, 2,2-dimethylbutane

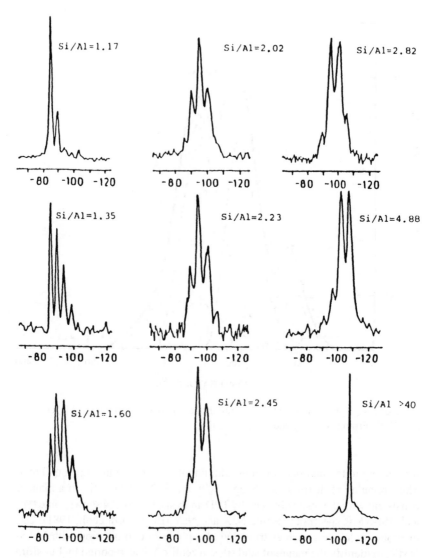

Figure 2.4 ^{29}Si NMR spectra of NaX and NaY zeolites. (From Englehardt et al., 1981.)

has a nearly circular cross section; benzene has an elliptical cross section), the shape of the pore, circular or elliptical, may be differentiated (Wu et al., 1986; Chester et al., 1987).

Derouane and Vanderveken (1988) showed the exceptional selectivity of Ba- and K-exchanged Pt-containing Linde Type L in the aromatization of *n*-hexane to benzene is related to the "confinement" catalysis. Thus, an integration of atomic chemical modification with molecular graphics and theoretical computations will provide a design of molecular sieve catalysts.

The use of catalytic diagnostic tests using model compounds of different sizes and shapes to characterize the pore size of zeolites, suggested earlier by Miale et al. (1966), is now being accepted as a valuable tool complementing the other physicochemical techniques. For example, the relative rate of cracking of *n*-hexane and 3-methylpentane was used to differentiate members of the erionite-offretite family (Chen et al., 1984) and the clinoptilolite-heulandite family of zeolites (Chen et al., 1978a).

A similar test known as the "constraint index" test (Frilette et al., 1981) has been devised to distinguish medium pore zeolites from large pore and small pore zeolites (Chen et al., 1978a; Dewing, 1984). Weitkamp and Ernst (1994) recently reviewed a series of experimental tests. These tests suggested the use of bulkier molecules, in addition to alkanes or mononuclear aromatics, as the probe chemicals, for example, naphthalene (Weitkamp and Neuber, 1991; Katayama et al., 1991) or biphenyl compounds (Lee et al., 1989; Butruille and Pinnavaia, 1992) for probing very large pore materials.

That acid catalyzed shape selective reactions can be achieved over small pore zeolites with pore openings comprising 8-membered oxygen rings (~5 Å in diameter) demonstrates that the acid activity must have originated within the intracrystalline cavities.

With low SiO_2/Al_2O_3 ratio large pore zeolites, the task of correlating various acid sites with their catalytic properties is made most difficult not only by the multiplicity of acid sites, but also by the added complexity of rapid coking and heat and mass transport effects due to their high acid site density and pore structure. Obviously this is an important task and numerous studies have dealt with this subject. However, the state of our knowledge relating catalytic properties to other characterizing information, particularly with respect to the relevance of fundamental understanding to practical applications, remains inadequate (see, for example, Barthomeuf and Beaumont, 1973; Barthomeuf, 1979, 1980; Jacobs, 1981; DeCanio et al., 1986; Sohn et al., 1986).

The quantitation of protonic sites in medium pore zeolites, which generally have much higher SiO_2/Al_2O_3 ratios than the large pore zeolites, appears at first glance to be less complicated.

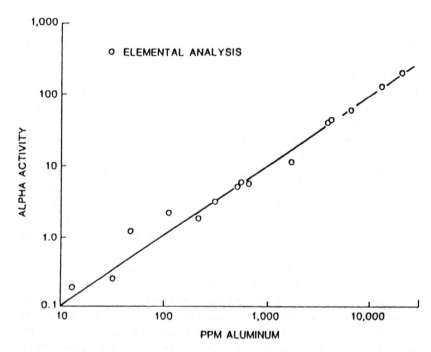

Figure 2.5 Relationship between acid activity and aluminum content. (From Olson et al., 1980.)

Using *n*-hexane (the alpha (α) test)* and *n*-hexene as the probing molecules and HZSM-5 samples carefully deammoniated from NH₄ZSM-5 in the absence of water vapor, an essentially linear relationship between catalytic activity and the number of protonic sites in HZSM-5 was established over a range of silica-to-alumina ratios from 35 to 60,000 (~15 ppm Al). As shown in Figure 2.5, this linear relationship also extrapolates to zero activity at zero aluminum content (Olson et al., 1980). Similar relationships have been reported

---

*The α test measures the intrinsic acid activity of a given sample on a relative basis by using *n*-hexane as the probe molecule (Weisz and Miale, 1965).

The "α value" of a catalyst is defined as the ratio of the first order rate constant of the sample to that of an arbitrary standard, the 46 AI (activity index) amorphous silica-alumina catalyst measured at 538°C. Thus, a sample has an α value of one when its activity at 538°C is the same as that of the standard. For most samples, measurements made at other than 538°C can be extrapolated to 538°C using a temperature correction factor corresponding to an apparent activation energy of 30 kcal/mol. However, for some small pore and medium pore zeolites, such as erionite and ZSM-5 which have different apparent activation energies, caution must be taken in making proper extrapolations.

for a number of acid catalyzed reactions including the disproportionation of toluene, the conversion of methanol to hydrocarbons (Haag, 1984), ethylbenzene dealkylation, and cyclopropane isomerization (Chu et al., 1985). As shown in Figure 2.6, excellent linear relationships have also been found to exist between the elemental analysis, TPAD, and Al MAS-NMR (Haag et al., 1984; Derouane et al., 1985). Results of using other techniques, including infrared analysis (Topsoe et al., 1981; Jacobs and von Ballmoos, 1982), ion exchange capacity, and reconstitution of NH_4^+ zeolite with ammonia (Beaumont and Barthomeuf, 1972; Jacobs, 1981), are consistent with the conclusion based on catalytic measurements that each framework Al atom associated with a proton gives rise to one Brønsted acid site.

IV. DISTRIBUTION AND LOCATION OF ACID SITES

Medium pore zeolites may be synthesized with only a trace concentration of aluminum. There is much speculation and uncertainty with respect to whether the placement of aluminum in the silica framework occurs by a random or ordered process, or whether aluminum atoms are located only in certain specific sites. Since the concentration of aluminum on the average is no more than one in ten of the tetrahedral framework positions, or a maximum of about 10% of the 96 Si + Al tetrahedrals per unit cell, the uncertainty is difficult to resolve.

Jacobs and von Ballmoos (1982) suggested that the acid sites are located at the channel intersections. Structurally speaking, every tetrahedral site is part of the channel intersections, although not all the oxygen atoms are located in the intersections. Therefore, the specific location of the protonic sites could conceivably affect its catalytic activity. This would be particularly pronounced in the case of bulky molecules. Derouane and co-workers (Derouane and Vedrine, 1980; Dejaifve et al., 1980) also speculated that the channel intersections are probably the loci of catalytic activity of ZSM-5. But so far, this remains a largely unresolved issue.

von Ballmoos and Meier (1981) also reported the presence of a concentration gradient of aluminum within a crystal. Higher aluminum concentration was found near the surface of the crystal relative to its bulk concentration (von Ballmoos and Meier, 1981; Gabelica et al., 1984a). While it is still debatable whether these concentration gradients are present in crystals of any size (Suib et al., 1980; Derouane et al., 1981), they are easily detectable in large crystals. Various synthesis methods have been designed to eliminate such gradients by maintaining a constant aluminum concentration during synthesis (Chen et al., 1978b) or to accentuate this gradient to produce a silicon-rich outer shell or an aluminum-rich shell by changing the gel composition abruptly during synthesis

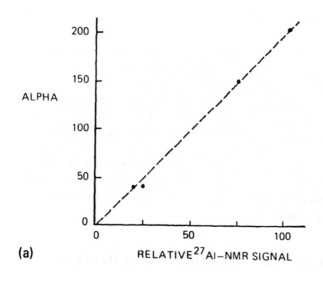

(a)

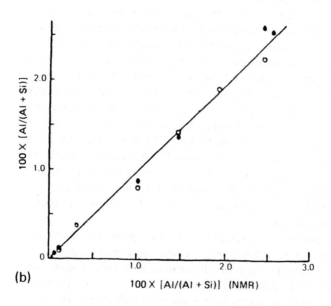

(b)

Figure 2.6 Relationships between catalytic activity, elemental analysis, and Al MAS-NMR. (a) Catalytic activity (alpha) vs. relative ^{27}Al-NMR signal. (b) Atomic fraction of framework Al from elemental analysis (●) or TPAD (○) vs. atomic fraction of framework Al from ^{27}Al MAS-NMR. (From Haag et al., 1984; Derouane et al., 1985.)

(Rollmann, 1978, 1979b; Miller, 1983; Koetsier, 1985). Thus, it is possible to synthesize zeolites possessing different spatially distributed acid sites. Similar concentration gradients can be expected when the zeolite is chemically modified to increase or decrease its framework aluminum concentration.

At least in principle, the spatial distribution of aluminum in the zeolite could be an important factor governing its effectiveness as a catalyst, since intracrystalline diffusion plays a major role in many reactions. This could be even more important in determining the shape selectivity of medium pore zeolite catalysts, because the degree of selectivity achievable could depend not only on the relative rate of diffusion of the feed and product molecules but also on the size of the crystals and the radial distribution of acid sites (Wei, 1982).

The contribution of the external surface of a zeolite crystal to its overall catalytic activity has often been questioned. With crystals larger than 1 micron, their external surface area relative to their intracrystalline surface area is so small ($<1\%$) that surface activity can be ignored. However, for small submicron crystals, external surface sites could be a significant fraction of the total surface area. If the external surface sites are catalytically either the same or more active than the intracrystalline sites, then the shape selectivity of a zeolite could conceivably be changed by these surface sites (Gilson and Derouane, 1984), an effect known as the "nest effect." One such speculation was advanced by Fraenkel et al. (1984), who proposed that acid sites located in the "half cavities" on the external crystal surface of ZSM-5 could be active for a second type of shape selective reaction. These sites were speculated to be responsible for the formation of such molecules as o-, m-xylene and 1,2,4,5-tetramethylbenzene (durene) in methanol conversion and the selective formation of such molecules as 2,6- or 2,7-dimethylnaphthalene in the methylation of naphthalene.

However, the data as presented appear insufficient to support this proposal because in the absence of information on diffusion coefficients and reaction rates, one cannot rule out the effect of diffusional constraints on product selectivity. Furthermore, the surface sites are probably less active than tetrahedrally coordinated internal sites.

V. STRUCTURAL EFFECT ON ACID SITES

A. Constraint Index

As acid catalysts, the medium pore zeolites have exceptional shape selectivity and low coking tendencies (Chen and Garwood, 1978; Walsh and Rollmann, 1979). As mentioned previously, a useful diagnostic test known as the "constraint index" can be used to characterize these medium pore zeolites (Frilette et al., 1981). The numerical value determined by this test is approximately the ratio of the cracking rate constants for *n*-hexane and 3-methylpentane. Medium pore zeolites usually have constraint indices in the approximate range of 3 to 12; zeolites with fairly large pores may have constraint indices of 1 to 3, and zeolites with a constraint index of less than 1 are regarded to be large pore zeolites. The constraint index serves as a useful measure of the extent to which a zeolite provides control of access to molecules of varying sizes to its internal structure and is generally used in Mobil's patent literature involved with shape selective catalysis. An example is shown in Table 2.3.

The observed values of the constraint index are largely independent of the crystal size of the zeolite. Therefore, their selectivity cannot be attributed to diffusion constraints of the reactant molecules. It is attributed to the steric constraint of the reaction intermediates inside the zeolite pores—a new type of

Table 2.3   Constraints Index (CI) of Typical Zeolites

Zeolite	CI	Test Temperature
ZSM-34	50	316
Erionite	38	316
ZSM-23	9.1	427
ZSM-11	5–8.7	371–316
ZSM-5	6–8.3	371–316
ZSM-22	7.3	427
ZSM-35	4.5	454
TMA-offretite	3.7	316
ZSM-48	3.5	538
Clinoptilotite	3.4	510
ZSM-12	2.3	316
ZSM-50	2.1	427
ZSM-38	2	510
MCM-22	1.5	454
Zeolite beta	0.6–2	316–399
ZSM-20	0.5	371
Omega (ZSM-4)	0.5	316
Mordenite	0.5	316
Dealuminized Y	0.5	316
TEA-mordenite	0.4	316
REY	0.4	316

Source: Chang and Hellring (1994).

shape selectivity, known as *spatiospecificity*, or *transition state selectivity* (Haag et al., 1981), which will be discussed in more detail in Chapter 3.

This interpretation is consistent with the classical mechanism of acid catalyzed paraffin cracking involving bimolecular hydride transfer as the rate-determining step. However, for a number of years there was no adequate explanation for the observation that the constraint index of medium pore zeolites decreases with increasing temperates (Chen and Garwood, 1978; Frilette, 1981), as shown in Figure 2.7.

Haag and co-workers (1984) demonstrated that the restricted pore geometry of the medium pore zeolites promoted a second paraffin cracking mechanism involving the direct protonation of a paraffin molecule. The reaction is monomolecular and therefore does not involve a bulky reaction intermediate as does the classical mechanism. It is reasonable to expect that in the absence of steric constraint, *n*-hexane and 3-methylpentane should crack at comparable rates by this mechanism. The increasing importance of this reaction at higher temperatures provides a plausible explanation of this unusual temperature dependency of the constraint index of medium pore zeolites.

Because the cracked products are different for the reaction pathway of these two cracking mechanisms, it is possible to estimate from the product distribution the contribution from each pathway.

B. Other Catalytic Characterization Tests

Characterization of Hydrogen Transfer Activity

Lukyanov (1994) proposed that a quantitative characterization of the hydrogen transfer activity of zeolites using *n*-hexane cracking at 400°C and determining the relative formation rate of isobutane and isobutene. Table 2.4 compares hydrogen transfer activity and alpha activity of HZSM-5 and HY.

A similar test known as the hydrogen transfer index test was proposed by Miller (1987) using 1-hexene in nitrogen at 210°C and 34 atm of pressure to determine the ratio of 3-methylpentene to 3-methylpentane at 30 to 70% conversion. Hydrogen transfer, a bimolecular reaction, is promoted in large pore zeolites. Table 2.5 shows the hydrogen transfer index for a number of zeolites at 50% 1-hexene conversion.

m-Xylene Isomerization and Disproportionation

Gnep et al. (1982) and Dewing (1984) used *m*-xylene conversion as the test reaction. Dewing used the ratio of *p*-xylene to *o*-xylene in *m*-xylene isomerization to characterize the pore size of various medium pore zeolites. Gnep measured not only the rate of isomerization, but also the relative rate of disproportionation to that of isomerization. More detailed studies of this test by

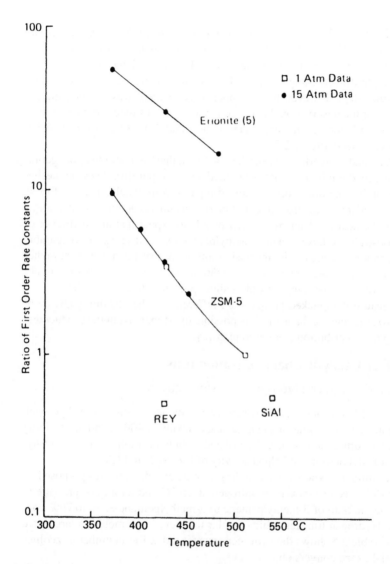

Figure 2.7 Shape selectivity between *n*-hexane and 3-methyl-pentane. (From Chen and Garwood, 1978.)

Table 2.4 Hydrogen Transfer Activity and Alpha Activity of HZSM-5 and HY

Type of activity	Relative activity	
	ZSM-5	HY
Hydrogen transfer activity	1	6.1
Alpha activity	1	0.36

Source: Lukyanov (1994).

Table 2.5 Hydrogen Transfer Index for a Number of Zeolites

Zeolite	Hydrogen transfter index
ZSM-5	55.0
Mordenite	8.0
Ultrastable Y	0.2

Source: Miller (1987).

Martens et al. (1988) and Kumar et al. (1989) indicated that the test is useful in distinguishing medium pore zeolites but has limited value for the characterization of large pore zeolites.

Isomerization and Transalkylation of 1-Ethyl-2-Methylbenzene

Csicsery (1987) applied this test to a broader variety of zeolites, including ZSM-5, offretite, mordenite, and beta and analyzed the product distribution. This test should be particularly applicable to distinguish medium pore and large pore zeolites.

Hydroisomerization and Hydrocracking of Long Chain Alkanes (Modified Constraint Index, CI*)

Hydroisomerization of long chain paraffins (C_9 to C_{16} paraffins) was studied extensively by Jacobs and Weitkamp and their co-workers (Jacobs et al., 1981; Weitkamp et al., 1983). Compared to large pore zeolites, such as the ultrastable Y (Jacobs et al., 1980), the channel structure of ZSM-5 was found to exert a pronounced effect on both the relative rate of cracking to isomerization and the isomer distribution (Jacobs et al., 1980, 1981).

Table 2.6 SI of Number of
Zeolites

Zeolite	SI
ZSM-22	1
ZSM-23	1
ZSM-5	1
ZSM-12	3
SAPO-5	4
EU-1	5
Offretite	5
Mordenite	7
L	17
Beta	18
ZSM-20	20
Y	21

Source: Weitkamp et al. (1988, 1989).

The shape selectivity effect (Martens et al., 1984) in the hydroisomerization of n-decane is defined as the modified constraint index, CI*, as

$$CI^* = \frac{Y_{2\text{-methylnonane}}}{Y_{5\text{-methylnonane}}} \quad \text{at } Y_{\text{isodecanes}} \approx 5\%$$

where Y is the yield

Hydrocracking of C_{10} Naphthenes (Spaciousness Index, SI)

In the hydrocracking of butylcyclohexane, the yield ratio of isobutane and n-butane (Weitkamp et al., 1986) was found to increase with increasing space inside the pores of the zeolite. This ratio was named the spaciousness index (SI). Table 2.6 shows the SI of a number of zeolites. This test seems to ideally suited for exploring the space available in the large variety of 12-membered oxygen ring zeolites.

C. Para Selectivity

One of the unique shape selective characteristics of HZSM-5 is its para selectivity (Chen, 1977, 1978; Chen et al., 1979) in electrophilic substitution reactions such as alkylation and disproportionation of alkyl aromatics.

By adjusting the acid activity of the zeolite and controlling the diffusion parameter, high para selectivity can be achieved. Generally, a necessary condition for good para selectivity is that:

$$\frac{1}{k} << \frac{R^2}{D}$$

where k is the reaction rate constant, R is the crystal radius, and D is the diffusivity of the species with the slowest diffusion rate.

Without changing the intrinsic activity of the catalyst and the size of the crystal, the effective diffusion characteristics of the catalyst can be altered by a number of techniques, including surface silylation (Chen et al., 1979); surface coking (Kaeding et al., 1984; Ashton et al., 1986); surface poisoning with bulky heterocyclic compounds (Chen, 1988); or impregnation with magnesium oxide (Kaeding et al., 1981a; Olson and Haag, 1984; Sato et al., 1985), antimony oxide (Butter, 1977), or phosphorus and boron compounds (Kaeding et al., 1981b, 1984).

Haag and co-workers (Weisz, 1980; Olson and Haag, 1984) have developed a quantitative model relating the observed para selectivity with acid activity and diffusion parameters for a large variety of modified and unmodified ZSM-5 catalysts.

VI. ENHANCED ACID SITES

While the characterization of protonic sites in the hydrogen form of siliceous medium pore zeolites appears straightforward, a general description of acid sites in zeolites is much more complex because hydrogen zeolites easily undergo thermal and hydrothermal reactions such as dehydroxylation, dealumination, and stabilization. These reactions can alter the nature of the acid sites (McDaniel and Maher, 1976; Poutsma, 1976).

It is well known that severe steaming of hydrogen zeolites reduces the number of framework tetrahedral aluminum sites and the catalytic acid activity (Chen and Smith, 1976); however, steaming at mild temperatures and under low partial pressure of water vapor and/or in the presence of ammonia can produce catalysts of higher activity than the parent zeolite. For example, Haag and co-workers (Lago et al., 1986) showed that by steaming in the presence of ammonia, the catalytic activity of a HZSM-5 was increased by a factor of 9. The increase in catalytic activity can be explained by the presence of a new type of acid site of higher specific activity. It is estimated by the α test that the intrinsic specific activity of these enhanced sites may be at least 40 times higher than that of the conventional protonic sites in ZSM-5. However, the nature of these enhanced sites and the chemistry of their formation have not been completely elucidated.

Equally interesting new ways of enhancing acidity by inserting aluminum into high-silica zeolite frameworks have been reported in the literature. Shihabi et al. (1985) showed that binding of high-silica HZSM-5 with alumina enhances the catalytic activity of the catalyst for numerous reactions and postulated a mechanism for the insertion of aluminum from the binder into the tetrahedral framework of the zeolite.

Alumination of high-silica ZSM-5 has also been achieved with aluminum chloride (Anderson et al., 1984; Dessau and Kerr, 1984) and with a variety of aluminum halides or aqueous ammonium fluoroaluminate (Chang et al., 1984; Miale and Chang, 1984). At least some of the added aluminum is shown to be incorporated into the zeolite as a tetrahedrally coordinated species by independent physical measurements such as ^{27}Al MAS-NMR, NH_3-TPD, and FTIR. The treated catalysts show increased acid activity in methanol conversion and paraffin cracking reactions and give products similar to those of conventional HZSM-5 catalysts (Chang et al., 1984). It remains unclear, however, whether these newly created acid sites have the same or different specific activity as the conventional protonic sites or the enhanced sites.

VII. ACID SITES IN ISOMORPHOUS SUBSTITUTED ZEOLITES

The possibility of creating acid sites by isomorphous substitution of silicon with other elements, such as berylium, boron, chromium, gallium, germanium, iron, phosphorus, and titanium, in addition to aluminum, either by direct synthesis or by chemical modification, is now known.

It is known that a number of natural minerals have the same framework topology but different chemical compositions (Barrer, 1982, 1984). Early claims of making isomorphous substituted chromium zeolites either by the synthesis route (Ermolenko et al., 1964) or by posttreatments (Garwood et al., 1978) were made, however, positive identification of framework substitution was difficult. Taramasso et al. (1980) and Klotz (1981) reported that by X-ray analyses of the unit cell volume of borosilicates, a method was devised to determine the framework boron content of a number of boron-substituted zeolites, including ZSM-5 and ZSM-11. The substitution of boron for silicon in ZSM-5 was further confirmed by ^{11}B MAS-NMR (Derouane et al., 1985) The insertion of boron was also accomplished using a Pyrex reaction vessel (Gabelica et al., 1984b) and by impregnation (Kaeding et al., 1981b; 1984; Gabelica et al., 1984a; Coudurier and Vedrine, 1986).

Available data on boron-substituted zeolites (Taramasso et al., 1980; Hölderich et al., 1984; Chu and Chang, 1985; Chang et al., 1985; Chu et al., 1985; Coudurier and Vedrine, 1986; Coudurier et al., 1987) indicate that the acid activity of the tetrahedrally coordinated framework boron sites is much lower than that of aluminum. In fact, Chu et al. (1985) attributed the observed catalytic acidity, if any, of B-ZSM-5 to trace amounts of aluminum sites present in the sample.

The synthesis of titanium-containing zeolites opened the field of epoxidation of alkanes and alkenes with hydrogen peroxide or *tert*-butyl hydroperoxide (Taramasso et al., 1980). More will be discussed in Chapter 4.

The introduction of gallium in zeolites has received considerable interest for the conversion of light paraffins to aromatics (Kitagawa et al., 1986; Price and Kanazirev, 1990; Abdul Hamid et al., 1994; Kwak et al., 1994). As will be discussed in Chapter 6, a cyclic aromatization processes, the Cyclar™ process, was announced by UOP and British Petroleum in 1984, converting LPG to aromatics used as high-octane gasoline blending components (Johnson and Hilder, 1984). Despite some theoretical disagreement on the active site of Ga, it is reasonable to believe that framework-substituted gallium is not an important factor contributing to the observed catalytic activity.

REFERENCES

Abdul Hamid, S. B., E. G. Derouane, G. Demortier, J. Riga, and M. A. Yarmo, Appl. Catal. *A108*, 83 (1994).

Absil, R. P. L., S. Han, D. S. Shihabi, J. C. Vartuli, and P. Varghese, U.S. Pat. 5,030,787, Jul. 9, 1991.

Anderson, M. W., J. Klinowski, and X. Liu, J. Chem. Soc., Chem. Commun. 1596 (1984).

Argauer, R. J. and G. R. Landolt, U.S. Pat. 3,702,886, Nov. 14, 1972.

Ashton, A. G., S. Batmanian, J. Dwyer, I. S. Elliott, and F. R. Fitch, J. Mol. Catal. *34*, 73 (1986).

Baerlocher, Ch., Acta Cryst. *A46*, C177 (1990).

Barrer, R. M., Chem. Ind. (London) 1203 (1968).

Barrer, R. M., *Hydrothermal Chemistry of Zeolites*, Academic Press, London, 1982.

Barrer, R. M, Proc. 6th Int. Zeolite Conf., D. Olson and Attilio Basio, eds., Butterworths, Surrey, UK, p. 870 (1984).

Barrer, R. M. and H. Villiger, Z. Kristallogr. *128*, 352 (1969).

Barri, S. A. I., G. W. Smith, D. White, and D. Young, Nature *312*, 533 (1984).

Barthomeuf, D., J. Phys. Chem. *83*, 249 (1979).

Barthomeuf, D., Stud. Surf. Sci. Catal. *5*, 55 (1980).

Barthomeuf, D. and R. Beaumont, J. Catal. *30*, 288 (1973).

Beagley, B., J. Dwyer, F. R. Fitch, R. Mann, and J. Walters, J. Phys. Chem. *88*, 1744 (1984).

Beaumont, R. and D. Barthomeuf, J. Catal. *26*, 218 (1972).

Beck, J. S., J. C. Vartuli, W. J. Roth, M. E. Leonowicz, C. T. Kresge, K. D. Schmitt, and C. T.-W. Chu, J. Am. Chem. Soc. *114*, 10834 (1992).

Bennett, J. M. and J. A. Gard, Nature *214*, 1005 (1967).

Bezak, J. M. and R. Mostowicz, Stud. Surf. Sci. Catal. *24*, 47 (1985).

Breck, D. W., *Zeolite Molecular Sieves*, Wiley, New York (1974).

Butruille, J. R. and T. J. Pinnavaia, Catal. Lett. *12*, 187 (1992).

Butter, S. A., U.S. Pat. 4,007,231, Feb. 8, 1977.

Chang, C. D. and S. D. Hellring, U.S. Pat. 5,288,927, Feb. 22, 1994.

Chang, C. D., C. T.-W. Chu, J. N. Miale, R. F. Bridger, and R. B. Calvert, J. Am. Chem. Soc. *106*, 8143 (1984).

Chang, C. D., S. D. Hellring, J. N. Miale, and K. D. Schmitt, J. Chem. Soc., Faraday Trans. 1, *81*, 2215 (1985).

Chen, N. Y., U.S. Pat. 4,002,697, Jan. 11, 1977.

Chen, N. Y., U.S. Pat. 4,100,215, Jul. 11, 1978.

Chen, N. Y., J. Catal. *114*, 17 (1988).

Chen, N. Y. and F. A. Smith, Inorg. Chem., *15*, 295 (1976).

Chen, N. Y. and W. E. Garwood, J. Catal. *52*, 453 (1978).

Chen, N. Y., W. J. Reagan, G. T. Kokotailo, and L. P. Childs, *Natural Zeolites*, L. B. Sand and F. A. Mumpton, eds., Pergamon, New York, p. 411 (1978a).

Chen, N. Y., J. N. Miale, and W. J. Reagan, U.S. Patent 4,112,056, Sept. 15, 1978b.

Chen, N. Y., W. W. Kaeding, and F. G. Dwyer, J. Am. Chem. Soc. *101*, 6783 (1979).

Chen, N. Y., J. L. Schlenker, W. E. Garwood, and G. T. Kokotailo, J. Catal. *86*, 24 (1984).

Chen, N. Y., W. E. Garwood, and S. B. McCullen, U.S. Pat. 4,814,543, Mar. 21, 1989.

Chen, N. Y., T. F. Degnan, Jr., C. R. Kennedy, A. B. Ketkar, L. R. Koenig, R. A. Ware, U.S. Pat. 4,911,823, Mar. 27, 1990.

Chester, A. W., P. Chu, and W. J. Rohrbaugh, "Relationships Between Measuring Zeolite Pore Sizes and Catalytic Performance," paper presented at the 10th North American Catal. Soc. Mtg., San Diego, CA, May 17–22, 1987.

Chu, C. T.-W. and C. D. Chang, J. Phys. Chem. *89*, 1569 (1985).

Chu, C. T.-W., G. H. Kuehl, R. M. Lago, and C. D. Chang, J. Catal. *93*, 451 (1985).

Chu, C. T.-W., T. F. Degnan, and B. K. Huh, U.S. Pat. 4,982,033, Jan. 1, 1991.

Coudurier, G. and J. C. Vedrine, Pure Appl. Chem. *58*, 1389 (1986).

Coudurier, G. A. Auroux, J. C. Vedrine, R. D. Farlee, L. Abrams, and R. D. Shannon, J. Catal. *108*, 1 (1987).

Csicsery, S. M., J. Catal. *108*, 433 (1987).

Davis, M. E., C. Montes, and J. M. Garces, Am. Chem. Soc. Symp. Ser. *398*, 291 (1989).

DeCanio, S. J., J. R. Sohn, P. O. Fritz, and J. H. Lunsford, J. Catal. *101*, 132 (1986).

Dejaifve, P., J. C. Vedrine, and E. G. Derouane, J. Catal. *63*, 331 (1980).

Derouane, E. G., Stud. Surf. Sci. Catal. *20*, 221 (1985).

Derouane, E. G. and D. J. Vanderveken, Appl. Catal. *45*, L15 (1988).

Derouane, E. G. and J. C. Vedrine, J. Mol. Catal. *8*, 479 (1980).

Derouane, E. G., J. P. Gilson, Z. Gabelica, C. Mousty-Desbuquoit, and J. Verbist, J. Catal. *71*, 447 (1981).

Derouane, E. G., L. Baltusis, R. M. Dessau, and K. D. Schmitt, Stud. Surf. Sci. Catal. *20*, 135 (1985).

Dessau, R. M. and G. T. Kerr, Zeolites *4*, 315 (1984).

Dessau, R. M., K. D. Schmitt, G. T. Kerr, G. L. Woolery, and L. B. Alemany, J. Catal. *104*, 484 (1987).

Dewing, J., J. Mol. Catal. *27*, 25 (1984).

Dima, E. and L. V. C. Rees, Zeolites *10*, 8 (1990).

Dwyer, F. G. and E. E. Jenkins, U.S. Pat. 3,941,871, Mar. 2, 1976.

Engelhardt, G., U. Lohse, E. Lippmaa, M. Tarmak, and M. Z. Magi, Anorg. Allg. Chem. *482*, 49 (1981).

Ermolenko, N. F., S. A. Levina, and L. V. Pansevitch-Kolada, Dokl. Akad. Nauk B. SSR *8(6)*, 394 (1964).

Fegan, S. G. and B. M. Lowe, J. Chem. Soc., Chem. Commun. 437 (1984).

Flanigan, E. M., J. M. Bennett, R. W. Grose, J. P. Cohen, R. L. Patton, R. M. Kirchner, and J. V. Smith, Nature *272*, 840 (1978a).

Flanigan, E. M., J. M. Bennett, R. W. Grose, J. P. Cohen, R. L. Patton, R. M. Kirchner, and J. V. Smith, Nature *271*, 512 (1978b).

Fraenkel, D., M. Cherniavsky, and M. Levy, Proc. 8th Int. Congr. Catal. *4*, 545 (1984).

Frilette, V. J., W. O. Haag, and R. M. Lago, J. Catal. *67*, 218 (1981).

Fyfe, C. A., H. Gies, G. T. Kokotailo, B. Marler, and D. E. Cox, J. Phys. Chem. *94*, 3718 (1990).

Gabelica, Z., E. G. Derouane, and N. Blom, Am. Chem. Soc. Symp. Ser. *248*, 219 (1984a).

Gabelica, Z., G. Debras, and J. B. Nagy, Stud. Surf. Sci. Catal. *19*, 113 (1984b).

Garwood, W. E., S. J. Lucki, N. Y. Chen, and J. C. Bailar, Jr., Inorg. Chem. *17*, 610 (1978).

Gilson, J. P. and E. G. Derouane, J. Catal. *88*, 538 (1984).

Gnep, N. S., J. Tejada, and M. Guisnet, Bull. Soc. Chim. Fr., *1*, 5 (1982).

Gorring, R. L., J. Catal. *31*, 13 (1973).

Grose, R. W. and E. M. Flanigan, U.S. Pat. 4,257,885, Mar. 24, 1981.

Haag, W. O., Proc. 6th Int. Zeol. Conf., D. H. Olson, and A. Bisio, eds., Butterworths, Surrey, U.K., p. 466 (1984).

Haag, W. O. and R. M. Dessau, Proc. 8th Int. Congr. Catal. *2*, 305 (1984).

Haag, W. O., R. M. Lago, and P. B. Weisz, Chem. Soc. Faraday Disc. *72*, 317 (1981).

Haag, W. O., R. M. Lago, and P. B. Weisz, Nature *309*, 589 (1984).

Higgins, J. B., R. B. LaPierre, J. L. Schlenker, A. C. Rohrman, Jr., J. D. Wood, G. T. Kerr, and W. J. Rohrbaugh, Zeolites *8*, 446 (1988).

Hölderich, W., H. Eichhorn, R. Lehnert, L. Marosi, W. Mross, R. Reinke, W. Ruppel, and H. Schlimper, Proc. 6th Int. Zeol. Conf., D. H. Olson and A. Bisio, eds., Butterworths, Surrey, UK, p. 545 (1984).

Innes, R. A., S. L. Zones, and G. J. Nacamuli, U.S. Pat. 5,081,323, Jan. 14, 1992.

Jacobs, P. A., Catal. Rev.-Sci. Eng. *24*, 415 (1981).

Jacobs, P. A. and R. von Ballmoos, J. Phys. Chem. *86*, 3050 (1982).

Jacobs, P. A., J. B. Uytterhoeven., M. Steyns, G. Froment, and J. Weitkamp, Proc. 5th Int. Zeol. Conf., L. V. Rees, ed., Heyden, London, p. 607 (1980).

Jacobs, P. A., J. A. Martens, J. Weitkamp, and H. K. Beyer, Chem. Soc. Faraday Disc. *72*, 353 (1981).

Johnson, J. A. and G. K. Hilder, "Dehydrocyclodimerization, Converting LPG to Aromatics," paper presented at NPRA Annual Mtg., San Antonio, TX, March 25–27, 1984.

Kaeding, W. W., C. Chu, L. B. Young, B. Weinstein, and S. A. Butter, J. Catal. *67*, 159 (1981a).

Kaeding, W. W., C. Chu, L. B. Young, and S. A. Butter, J. Catal. *69*, 392 (1981b).

Kaeding, W. W., L. B. Young, and C. C. Chu, J. Catal. *89*, 267 (1984).

Kapustin, G. I., T. R. Brueva, A. L. Klyachko, S. Beran, and B. Wichterlova, Appl. Catal. *42*, 239 (1988).

Katayama, A., M. Toba, G. Takeuchi, F. Mizukami. S. Niwa, and S. Mitamura, J. Chem. Soc., Chem. Commun. 39 (1991).

Kerr, G. T., Catal. Rev.-Sci. Eng. *23*, 281 (1981).

Kibby, C. L., A. J. Perrotta, and F. E. Massoth, J. Catal. *35*, 256 (1974).

Kitagawa, H., Y. Sendoda, and Y. Ono, J. Catal. *101*, 12 (1986).

Klinowski, J., S. Ramdas, J. M. Thomas, C. A. Fyfe, and J. S. Hartman, J. Chem. Soc., Faraday Trans. 2, *78*, 1025 (1982).

Klotz, M. R., U.S. Pat. 4,268,420, May 19, 1981.

Koetsier, W., European Pat. EP 0 054 385 B1, Aug. 14, 1985; EP 0 055 044 B1, Sep. 18, 1985.

Kokotailo, G. T. and W. M. Meier, Chem. Soc., Spec. Publ. *33*, 133 (1980).

Kokotailo, G. T., S. L. Lawton, D. H. Olson, and W. M. Meier, Nature *272*, 437 (1978a).

Kokotailo, G. T., P. Chu, S. L. Lawton, and W. M. Meier, Nature *275*, 119 (1978b).

Kokotailo, G. T., J. L. Schlenker, F. G. Dwyer, and E. W. Valyocsik, Zeolites, *5*, 349 (1985).

Kresge, C. T., M. E. Leonowicz, W. J. Roth, J. C. Vartuli, and J. S. Beck, Nature *359*, 710 (1992).

Kumar, R., G. N. Rao, and P. Ratnasamy, Stud. Surf. Sci. Catal. *49B*, 1141 (1989).

Kwak, B. S., W. M. H. Sachtler, and W. O. Haag, J. Catal. *149*, 465 (1994).

Lago, R. M., W. O. Haag, R. J. Mikovsky, D. H. Olson, S. D. Hellring, K. D. Schmitt, and G. T. Kerr, Proc. 7th Int. Zeol. Conf., Y. Murakami, A. Iijima, and J. W. Ward eds., Kodansha/Elsevier, Tokyo/Amsterdam, p. 677 (1986).

La Pierre, R. B., R. D. Partridge, N. Y. Chen, and S. S. Wong, U. S. Pat. 4,419,220, Dec. 6, 1983.

LaPierre, R. B., A. C. Rohrman, Jr., J. L. Schlenker, J. D. Wood, M. K. Rubin, and W. J. Rohrbaugh, Zeolites *5*, 346 (1985).

Lee, G. S., J. J. Maj, S. C. Rocke, and J. M. Garcs, Catal. Lett. *2*, 243 (1989).

Leonowicz, M. E., J. A. Lawton, S. L. Lawton, and M. K. Rubin, Science *264*, 1910 (1994).

Lobo, R. F., M. Pan, I. Chan, H.-X. Li, R. C. Medrud, S. I. Zones, P. A. Crozier, and M. E. David, Science *262*, 1543 (1993).

Lobo, R. F., M. Pan, I. Chan, R. C. Medrud, S. I. Zones, P. A. Crozier, and M. E. Davis, J. Phys. Chem. *98*, 12040 (1994).

Lok, B. M., C. A. Messina, R. L. Patton, R. T. Gajek, T. R. Cannan, and E. M. Flanigan, U.S. Pat. 4,440,871, Apr. 3, 1984.

Lok, B. M., B. K. Marcus, and E. M. Flanigan, U.S. Pat. 4,500,651, Feb. 19, 1985.

Lukyanov, D.B., J. Catal. *145*, 54 (1994).

Lunsford, J. H., "Acid Sites in Zeolite Y: NMR Studies Using Probe Molecules," paper presented at the AIChE Nat. Mtg., New Orleans, Apr. 6, 1986.

Martens, J. A., M. Tielen, P. A. Jacobs, and J. Weitkamp, Zeolites *4*, 98 (1984).

Martens, J. A., J. Perez-Pariente, E. Sastre, A. Corma, and P. A. Jacobs, Appl. Catal. *45*, 85 (1988).

McDaniel, C. V. and P. K. Maher, *Zeolite Chemistry and Catalysis*, J. A. Rabo, ed., Am. Chem. Soc. Monogr. *171*, 316 (1976).

Meier, W. M. and D. H. Olson, *Atlas of Zeolite Structure Types*, Int. Zeolite Assoc., Butterworth-Heinemann, Boston, MA, 1992.

Meier, W. M., Z. Kristallogr. *115*, 439 (1979).

Miale, J. N. and C. D. Chang, U.S. Pats. 4,427,786 through 4,427,790, Jan. 24, 1984.

Miale, J. N., N. Y. Chen, and P. B. Weisz, J. Catal. *6*, 278 (1966).

Mikovsky, R. J. and J. F. Marshall, J. Catal. *44*, 170 (1976).

Mikovsky, R. J. and J. F. Marshall, J. Catal. *49*, 120 (1977).

Mikovsky, R. J., J. F. Marshall, and W. P. Burgess, J. Catal. *58*, 489 (1979).

Miller, S. J., U.S. Pat. 4,394,362, Jul. 19, 1983.

Miller, S. J., Stud. Surf. Sci. Catal. *38*, 187 (1987).

Miller, S. J., U.S. Pat. 4,859,311, Aug. 22, 1989.

Nastro, A., C. Colella, and R. Aiello, Stud. Surf. Sci. Catal. *24*, 39 (1985).

Olson, D. H. and W. O. Haag, Am. Chem. Soc. Symp. Ser. *248*, 275 (1984).

Olson, D. H., R. M. Lago, and W. O. Haag, J. Catal. *61*, 390 (1980).

Olson, D. H., G. T. Kokotailo, S. L. Lawton, and W. M. Meier, J. Phys. Chem. *85*, 2238 (1981).

Ozin, G. A., Advan. Mat. *4*, 612 (1992).

Plank, C. J., E. J. Rosinski, and M. K. Rubin, U.S. Pat. 4,016,245, Apr. 5, 1977.

Plank, C. J., E. J. Rosinski, and M. K. Rubin, U.S. Pat. 4,175,114, Nov 20, 1979.

Pluth, J. J. and J. V. Smith, Am. Mineral. *75*, 301 (1990).

Poutsma, M. L., "Zeolite Chemistry and Catalysis," J. A. Rabo, ed., Am. Chem. Soc. Monogr. *171*, 437 (1976).

Price, G. L. and V. Kanazirev, J. Catal. *126*, 267 (1990).

Ramdas, S., J. M. Thomas, J. Klinowski, C. A. Fyfe, and J. S. Hartman, Nature *292*, 228 (1981).

Rohrman, A. C., Jr., R. B. La Pierre, J. L. Schlenker, J. D. Wood, E. W. Valyocsik, M. K. Rubin, J. B. Higgins, and W. J. Rohrbaugh, Zeolites, *5*, 352 (1985).

Rollmann, L. D., U.S. Pat. 4,088,605, May 9, 1978.

Rollmann, L. D., Advan. Chem. Ser. *173*, 387 (1979a).

Rollmann, L. D., U.S. Pat. 4,148,713, Apr. 10, 1979b.

Rollmann, L. D. and D. E. Walsh, J. Catal. *56*, 139 (1979).

Rollmann, L. D. and D. E. Walsh, "Progress in Catalyst Deactivation," Nijhoff, The Hague, p. 81, 1982.

Rubin, M. K. and P. Chu, U.S. Pat. 4,954,325, Sep. 4, 1990.

Rubin, M. K., E. J. Rosinski and C. J. Plank, U. S. Pat. 4,086,186, Apr. 25, 1978.

Sachtler, J. W. A., R. J. Lawson, and S. L. Lambert, U.S. Pat. 5,081,084, Jan. 14, 1992.

Sato. H., N. Ishii, and S. Nakamura, U.S. Pat. 4,499,321, Feb. 12, 1985.

Sawa, M., M. Niwa, and Y. Murakami, Zeolites *11*, 93 (1991).

Schlenker, J. L., W. J. Rohrbaugh, P. Chu, E. W. Valyocsik, and G. T. Kokotailo, Zeolites *5*, 355 (1985).

Schlenker, J. L., J. B. Higgins, and E. W. Valyocsik, Zeolites *10*, 293 (1990).

Seddon, D. and T. V. Whittam, European Pat. B-55,529 (1985).

Shannon, M. D., J. L. Casci, P. A. Cox, and S. J., Nature *353*, 417 (1991).

Shihabi, D. S., W. E. Garwood, P. Chu, J. N. Miale, R. M. Lago, C. T. W. Chu, and C. D. Chang, J. Catal. *93*, 471 (1985).

Sohn, J. R., S. J. DeCanio, and J. H. Lunsford, J. Phys. Chem. *90*, 4847 (1986).

Suib, S. L., G. D. Stucky, and R. J. Blattner, J. Catal. *65*, 174 (1980).

Staples, L. W. and J. A. Gard, Min. Mag. *32*, 261 (1959).

Taramasso, M., G. Perego, and B. Notari, Proc. 5th Int. Zeol. Conf., L. V. Rees, ed., Heyden, London, p. 40 (1980).

Topsoe, N. Y., K. Pedersen, and E. G. Derouane, J. Catal. *70*, 41 (1981).

Vartuli, J. C., Kresge, C. T., W. J. Roth, S. B. McCullen, J. S. Beck, K. D. Schmitt, M. E. Leonowicz, J. D. Lutner, and E. W. Sheppard, Symp. Advan. Tech. in Catal. Prep., Am. Chem. Soc. Nat. Mtg., Apr. 1995.

Venuto, P. B., L. A. Hamilton, and P. S. Landis, J. Catal. *5*, 484 (1966).

Venuto, P. B. and P. S. Landis, Advan. Catal. *18*, 259 (1968).

von Ballmoos, R. and W. M. Meier, Nature *289*, 782 (1981).

von Ballmoos, R. and J. B. Higgins *Collection of Simulated XRD Powder Patterns for Zeolites*, Int. Zeolite Assoc. Butterworth, London, 1990.

Walsh, D. E. and L. D. Rollmann, J. Catal. *49*, 369 (1977).

Walsh, D. E. and L. D. Rollmann, J. Catal. *56*, 195 (1979).

Wang, F., W. Cheng, and S. Chang, J. Catal. (Chinese) *2*, 282 (1981).

Wei, J., J. Catal. *76*, 433 (1982).

Weisz, P. B. and J. N. Miale, J. Catal. *4*, 527 (1965).

Weisz, P. B., Proc. 7th Int. Congr. Catal. *1*, 3 (1980).

Weitkamp, J. and S. Ernst, Catal. Today *19*, 107 (1994).

Weitkamp, J. and M. Neuber, Stud. Surf. Sci. Catal. *60*, 291 (1991).

Weitkamp, J., P. A. Jacobs, and J. A. Martens, Appl. Catal. *8*, 123 (1983).

Weitkamp, J., S. Ernst, and R. Kumar, Appl. Catal. *27*, 207 (1986).

Weitkamp, J., C. Y. Chen, and S. Ernst Stud. Surf. Sci. Catal. *44*, 343 (1988).

Weitkamp, J., S. Ernst, and C. Y. Chen, Stud. Surf. Sci. Catal. *49*, 1115 (1989).

Wilson, S. T. and E. M. Flanigan, U.S. Patent 4,567,029, Jan. 28, 1986.

Wilson, S. T., B. M. Lok, and E. M. Flanigan, U.S. Patent 4,310,440, Jan. 12, 1982a.

Wilson, S. T., B. M. Lok, C. A. Messina, T. R. Cannan, and E. M. Flanigan, J. Am. Chem. Soc. *104*, 1146 (1982b).

Winquist, B. H. C., U.S. Pat. 3,933,974, Jan. 20, 1976.

Wu, E. L., G. R. Landolt, and A. W. Chester, Proc. 7th Int. Zeol. Conf., Y. Murakami, A. Iijima, and J. W. Ward, eds., Kodansha/Elsevier, Tokyo/Amsterdam, p. 547 (1986).

Yushchenko, V. V., A. N. Zakharov, and B. V. Romanovskii, Kinet. Katal. *27*, 474 (1986).

3

Principal Methods of Achieving Molecular Shape Selectivity

I. SIZE EXCLUSION

Separation among molecules of different sizes and shapes can be achieved by the proper choice of zeolites. The effective pore size of the zeolite can be further modified by the choice of cations associated with the zeolites. Catalytically, molecular shape selectivity by size exclusion can be achieved either through reactant selectivity or product selectivity. Reactant selectivity occurs when the feedstock contains two classes of molecules, one of which is too large to pass through the channel system of the zeolite. Product selectivity occurs when, among the multiplicity of products that could be formed, only those with the proper shape and size can pass through the channels as products.

   Shown in the following tables are some of the early examples demonstrating the principle of size exclusion in shape selective catalysis. They include the use of 8-ring zeolites such as zeolite A, T, erionite and chabazite and stacking faulted 12-ring zeolites such as gmelinite and mordenite. The selective dehydration of n-butanol without reacting isobutanol was demonstrated over the zeolite CaA (Table 3.1) (Weisz et al., 1962). Reactant selectivity was also demonstrated by the selective cracking of n-hexane in the presence of 3-

Table 3.1 Dehydration of Primary Butyl Alcohols (1 atm Pressure with 6-sec Residence Time)

Temperature, °C	130	230	260
Over Linde CaA molecular sieve			
isobutyl alcohol dehydrated, wt %		<2	<2
n-butyl alcohol dehydrated, wt %		18	60
sec-butyl alcohol dehydrated, wt %	~0		
Over faujasite-type CaX molecular sieve			
isobutyl alcohol dehydrated, wt %		46	85
n-butyl alcohol dehydrated, wt %		9	64
sec-butyl alcohol dehydrated, wt %	82		

Source: Weisz et al. (1962).

Table 3.2 Molecular Shape Selective Cracking

Catalyst	Hydrocarbon charge	Time on stream, min	Temperature, °C	Conversion, %
H Gmelinite	*n*-Hexane	10 to 33	370	47 to 30
H Gmelinite	2-Methylpentane	10 to 20	320 to 540	0 to 0.7
H Gmelinite	Methylcyclopentane	10 to 20	510 to 540	0.4 to 1.9
H Erionite	*n*-Hexane	26	320	52.1
H Erionite	2-Methylpentane	26	430	1.0
H Erionite	2-Methylpentane	26	540	4.7
H Chabazite	*n*-Hexane	30	260	10.0
H Chabazite	2-Methylpentane	10	540	1.5

Source: Miale et al. (1966).

methylpentane (Table 3.2) (Miale et al., 1966). Product selectivity was demonstrated by the near absence of branched chain products such as isobutane and isopentane (Table 3.3) (Weisz et al., 1962).

A more recent study was made by Heck and Chen (1992) in the hydrocracking of *n*-butane and *n*-heptane over a sulfided nickel erionite catalyst. This catalyst contained a weak dehydrogenation/hydrogenation function, which prevented the catalyst from rapid deactivation but did not participate as a bifunctional catalyst, and the distribution of "hydrocracked" products was similar to that for monofunctional acid catalysts.

Table 3.3 Comparison of *n*-Hexane and 3-Methylpentane Cracking at 500°C

Catalyst	3-Methyl pentane cracking conversion, %	*n*-Hexane cracking		
		Conversion, %	$\dfrac{\text{Iso-C}_4}{n\text{-C}_4}$	$\dfrac{\text{Iso-C}_5}{n\text{-C}_5}$
96% silica chips	<1	1.1		
Amorphous silica-alumina (46AI)	28	12.2	1.4	10
Linde NaA	<1	1.4		
Linde CaA	<1	9.2	<0.05	<0.05

Source: Weisz et al. (1962).

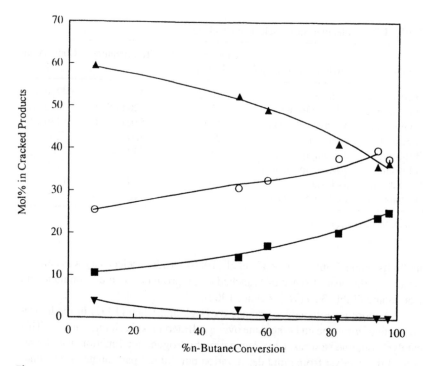

Figure 3.1 Products from *n*-butane cracking. (○) methane, (■) ethane, (▲) propane, (▼) pentane. (From Heck and Chen, 1992.)

The product distribution as shown in Figure 3.1, points to a reaction mechanism involving primary cracking, secondary cracking, and bimolecular reactions of the feed with itself and/or with primary or secondary reaction products.

Assuming that methane and ethane undergo no further reaction, the transformations used to fit the observed product distribution from *n*-butane, along with their relative rate constants determined by a finite difference fit and the resulting first-order differential equations, are shown in Table 3.4. It is important to note that although alkenes are not observable in the feed or products, they are present and play a role in monomolecular and bimolecular reactions.

It is seen from Table 3.4 that Transformation 2 is the fastest transformation of *n*-butane to C_3 and C_1. The secondary reactions represented by Transformations 3 and 4, involving hydrogen transfer, oligomerization, and cracking, make it possible to produce such a high propane yield, as indicated by Transformation 3, and the yield of C_5 from *n*-butane.

With medium pore zeolites, the cut-off point of molecular size exclusion is extended to include multiple branched aliphatic molecules and polyalkyl and polycyclic aromatics. However, for many molecules, the cut-off depends on temperature (see Section III on configurational diffusion). For example, at room temperatures, 1,3,5-trimethylbenzene (mesitylene) does not enter the pores of ZSM-5 (Olson et al., 1981; Wu and Landolt, 1986). Chang (1983) showed that in the conversion of methanol to hydrocarbons over ZSM-5, where methylation of aromatic rings is an important reaction, hexamethylbenzene is not formed, and there is a sharp cut-off point in product molecular weight. Among the aromatics made from methanol (Table 3.5), 1,3,5-trimethylbenzene

Table 3.4 Molecular Transformations Used to Fit *n*-Butane Selectivity

Number	Carbon balanced transformations	Relative rate constants (fit shown in Fig. 3.1)
1	$C_4 \rightarrow C_3 + C_1$	6.75
2	$C_4 \rightarrow 2C_2$	1.63
3	$C_4 \rightarrow 1.33\,C_3$	1.50
4	$C_4 \rightarrow 0.8\,C_5$	5.88
5	$C_3 \rightarrow C_2 + C_1$	2.00
6	$C_5 \rightarrow C_4 + C_1$	100.00

Source: Heck and Chen (1992).

Table 3.5 Aromatics Distribution from Methanol Conversion Over HZSM-5

	Normalized distribution, wt %	Normalized isomer distributions	Equilibrium distributions, 371°C
Benzene	4.1		
Toluene	25.6		
Ethylbenzene	1.9		
Xylenes:			
o	9.0	⎡21.5⎤	⎡23.8⎤
m	22.8	|54.6|	|52.7|
p	10.0	⎣23.9⎦	⎣23.5⎦
Trimethylbenzenes:			
123	0.9	⎡ 6.4⎤	⎡ 7.8⎤
124	11.1	|78.7|	|66.0|
135	2.1	⎣14.9⎦	⎣26.2⎦
Ethyltoluenes:			
o	0.7		
m + *p*	4.1		
Isopropylbenzene	0.2		
Tetramethylbenzenes:			
1234	0.4	⎡ 9.3⎤	⎡16.0⎤
1235	1.9	|44.2|	|50.6|
1245	2.0	⎣46.5⎦	⎣33.4⎦
Other A_{10}[a]	2.7		
$A_{11}{}^{+}$	0.4		

[a]Diethylbenzenes + dimethylethylbenzenes.
Source: Chang (1983).

and 1,2,3,4- and 1,2,3,5-tetramethylbenzenes fall far short of their equilibrium concentrations.

The isomerization and disproportionation of *m*-xylene was studied by Martens et al. (1988) and Kumar et al. (1989). Their distribution of trimethylbenzenes was summarized by Weitkamp and Ernst (1994) and is shown in Table 3.6. The data clearly show that as the pore size of the zeolites increases, the bulkier isomers increase toward the chemical equivalent values.

Similarly, in the hydroisomerization of paraffins (Table 3.7) tribranched isomers are not formed and less amounts of dibranched isomers are formed when compared with the large pore zeolite Y (Jacobs et al., 1981).

Table 3.6 Distribution of Trimethylbenzene Isomers Formed by Disproportionation of *m*-Xylene at 5–10% Conversion

Zeolite	$X_{1,2,3}$	$X_{1,2,4}$	$X_{1,3,5}$	$\dfrac{X_{1,2,3}}{X_{1,3,5}}$
ZSM-23	0	100	0	—
ZSM-5	0	100	0	—
ZSM-48	0	100	0	—
ZSM-50	7.7	84.6	7.7	1.0
ZSM-12	0	100	0	—
Offretite	4.7	95.3	0	
Mordenite	9.1	71.7	19.2	0.47
ZSM-4	9.0	77.4	13.6	0.66
Zeolite Beta	4.2	73.2	22.6	0.19
L	6.3	67.5	26.2	0.24
Y	7.9	64.0	28.1	0.26
Equilibrium @350°C	8.0	68.0	24.0	0.33

Source: Weitkamp and Ernst (1994).

Table 3.7 Distribution of the Feed Isomers from *n*-Decane over Pt/H-Zeolites at 50% Total Conversion

	Y*	ZSM-5	ZSM-11
Monobranched	65.0	95.0	85.0
Dibranched	30.0	5.0	15.0
Tribranched	5.0	—	—

Source: Jacobs et al. (1981).

II. COULOMBIC FIELD EFFECTS

The ionic character of the intracrystalline surface and the strength of coulombic field interaction between the zeolite and the sorbed molecules are functions of the silica/alumina ratio of the zeolite and the exchanged cations. Sodium exchanged low SiO_2/Al_2O_3 zeolites have a strong electrostatic field, while the hydrogen form of high SiO_2/Al_2O_3 zeolites do not. Thus the nature of the intracrystalline surface differs significantly among these zeolites, ranging from highly hydrophilic to substantially hydrophobic. Chen (1976) showed that as framework aluminum is removed from mordenite, the dealuminized zeolite becomes a hydrophobic sorbent. The degree of hydrophobicity is dependent on the SiO_2/Al_2O_3 ratio of the sample, suggesting that the adsorption of water involves a specific interaction with the tetrahedrally coordinated aluminum and associated cation centers. Figure 3.2 shows a linear relationship between the water sorbed and the concentration of alumina in the sample at a constant partial pressure of water. The slope of the straight lines corresponds to a coordination number of 4 water molecules per aluminum. Olson et al. (1980) found a similar linear relationship between the water sorption capacity of HZSM-5 and the alumina content of the samples. These hydrophobic zeolites retain their hydrocarbon sorption capacity. This is shown in Figure 3.3 for mordenite and ZSM-5 of different SiO_2/Al_2O_3 ratios (Chen, 1976; Nakamoto and Takahashi, 1982). Shown in Figure 3.4 are the absorption isotherms of water, *n*-hexane, methanol, and dichloromethane at 25°C for a 54/1 SiO_2/Al_2O_3 HZSM-5 (Chen, 1973). With these hydrophobic zeolites it is possible to separate polar compounds from less polar compounds, such as water and alcohols (Chen and Miale, 1983). Goldstein (1967) showed that acetic acid having a smaller van der Waals radius than *cis*-butene-2 is completely excluded by the zeolite CaA while the larger *cis*-butene-2 molecule is

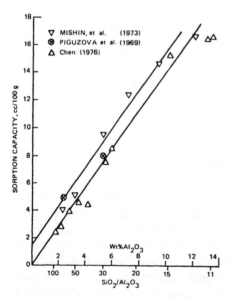

Figure 3.2   Water sorption at 25°C and 12 mm Hg of water. (From Chen, 1976.)

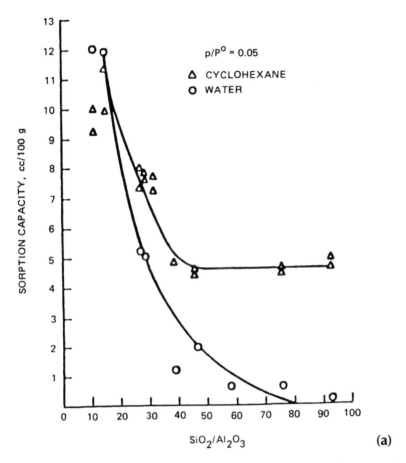

Figure 3.3 (a) Intracrystalline sorption at 25°C. (b) Amounts of water, methanol, benzene, and *n*-hexane absorbed on HZSM-5 at 100°C as a function of their SiO_2/Al_2O_3. (●) water, (▲) methanol, (□) *n*-hexane, (○) benzene. [From (a) Chen, 1976, and (b) Nakamoto and Tokahaski, 1982.]

sorbed. Thus a more sophisticated selectivity could be achieved beyond the realm of simple size exclusion.

Compared to amorphous silica/alumina, zeolites have been shown to have several orders of magnitude higher catalytic activity (Miale et al., 1966). This higher activity can be attributed to stronger coulombic field effect, which leads to higher reactant concentration inside the zeolites as compared to amorphous solids.

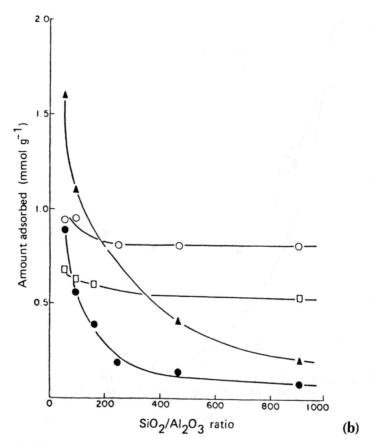

Figure 3.3 Continued.

III. CONFIGURATIONAL DIFFUSION

Configurational diffusion (Weisz, 1973) occurs in situations where the structural dimensions of the catalyst approach those of molecules. In this diffusion regime, even a subtle change in the dimensions of a molecule can result in a large change in its diffusivity, as shown in Figure 3.5.

For example, the diffusivity of *trans*-butene-2 is at least 200 times that of *cis*-butene-2 in zeolite CaA even though these two molecules differ in size by only about 0.2 Å (Chen and Weisz, 1967). Thus the rate of hydrogenation of

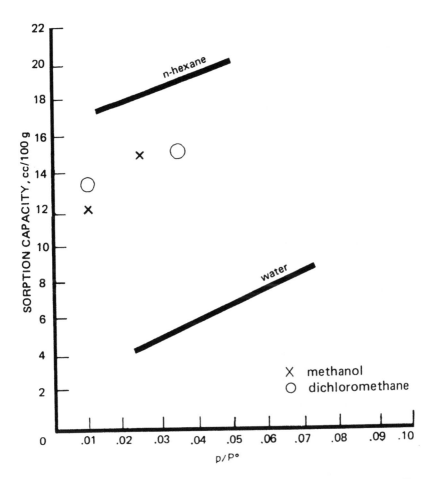

Figure 3.4   Absorption isotherms at 25°C, ZSM-5 (SiO_2/Al_2O_3 = 54). (From Chen, 1973.)

trans-butene-2 over Pt/zeolite A can be much faster than that of *cis*-butene-2, as shown in Table 3.8.

The favorable formation of *trans*-butene-2 in the double-bond isomerization of *n*-butenes over SAPO-11 was found by Richter et al. (1994). SAPO-11, similar in structure to ZSM-23, has 10 membered rings with elliptical aperture of 3.9 × 6.3 Å.

Tatsumi et al. (1991) showed that the epoxidation of a mixture of *trans/cis*-2-hexene over a titano-silicate (a ZSM-5 isomorph) preferentially produced the trans epoxide due to the faster rate of diffusion of the trans isomer.

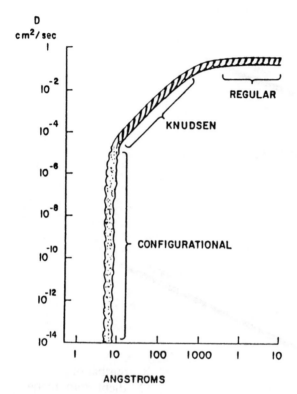

Figure 3.5 Configurational diffusion. (From Weisz, 1973.)

Table 3.8 Hydrogenation of a Mixture of *trans*- and *cis*-Butene-2

Temp., °C	Initial composition		Final composition			Conversion ϵ, wt %		$\dfrac{k_{\text{trans}}{}^{\text{a}}}{k_{\text{cis}}}$
	trans	cis	trans	cis	*n*-Butane	trans	cis	
120	78.7	21.3	37.2	17.0	45.9	52.9	10.8	3.3
103	78.7	21.3	57.3	19.8	22.9	27.2	7.1	4.3
98	78.7	21.3	69.4	20.9	9.7	11.8	1.8	7.0

[a]$k_{\text{trans}}/k_{\text{cis}} = \ln(1 - \epsilon_{\text{cis}})/\ln(1 - \epsilon_{\text{trans}})$.
Source: Chen and Weisz (1967).

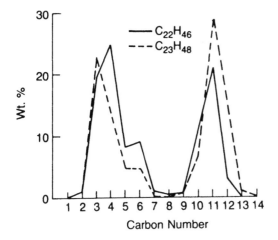

Figure 3.6 Carbon-number distribution of cracked products over erionite at 340°C. (From Chen et al., 1969.)

To achieve molecular shape selectivity, not only the size but also the dynamics of the molecular structure needs to be considered. For example, a window or cage effect was found with zeolite T (Gorring, 1973) and erionite (Chen et al., 1969) that imposes a strong but varying diffusional constraint on the diffusing paraffinic molecules of different chain lengths. As shown in Figure 3.6, when a n-C_{22} paraffin or a n-C_{23} paraffin is cracked over erionite, there are two distinct maxima in the size distribution of the product molecules and a distinct minimum at a carbon number of 8. This nonlinear chain length effect was also observed in hydrocracking n-paraffins of varying chain length (Figure 3.7) over erionite (Chen and Garwood, 1973) and ferrierite (Gianneti and Perrotta, 1975). Interestingly, as shown by Figure 3.8, the diffusivity of n-paraffins in zeolite T, which has a channel system similar to erionite, also changes in a similar periodic pattern by over two orders of magnitude between the minimum at C_8 and the maxima at C_4 and C_{11}.

The window, or cage, effect was hypothesized by Gorring: when the size of the zeolite cage matches the length of the n-paraffin molecules (in this case n-C_8H_{18}), the cage will trap the molecule and thus slow down its diffusivity. Molecules either smaller or larger than n-C_8H_{18} will not need to reorient themselves in the cage and will have higher diffusivity. Derouane et al. (1988a) explained this effect by the so-called surface curvature effect, i.e., the window effect is due to the variation in the sticking of the molecules to the

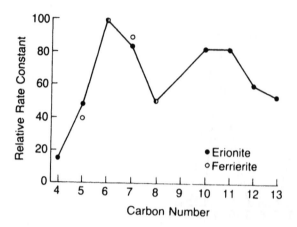

Figure 3.7 Nonlinear chain length effect in hydrocracking *n*-paraffins of varying chain length over erionite (Chen and Garwood, 1973) and ferrierite (Gianneti and Perrotta, 1975).

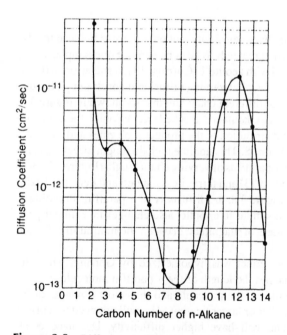

Figure 3.8 Diffusion coefficients of *n*-paraffins in potassium at 300°C. (From Gorring, 1973.)

pore wall rather than the difference in the ability of the molecules to reorient in the cages.

Nitsche and Wei (1991) published a theoretical analysis based on an analogy of the configuration diffusion process with an "equivalent" one-dimensional Brownian motion of a rod through a periodic sequence of potential barriers. Their numerical calculations as shown in Figure 3.9 are in reasonable agreement with Gorring's experimental data except that the hump at the C_4 carbon found by Gorring experimentally did not appear in the theoretical curve, while Derouane et al.'s surface curvature analysis showed such a hump.

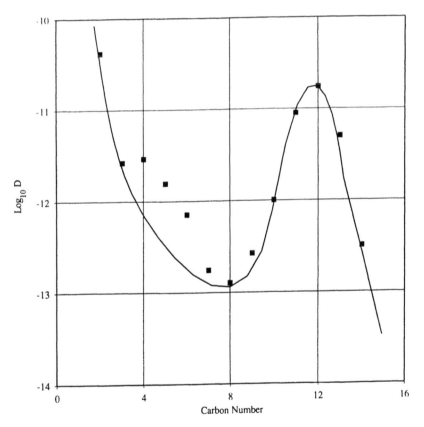

Figure 3.9 Diffusivities as a function of carbon number for diffusion of *n*-alkanes through zeolite T. (From Nitsche and Wei, 1991.)

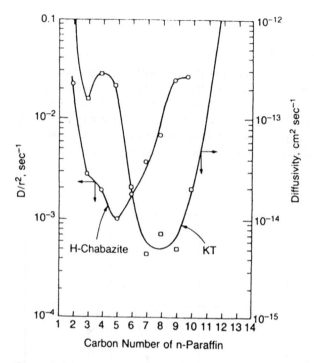

Figure 3.10 Diffusion of *n*-paraffins in H-chabazite KT. (From Gorring, R. L., private communication.)

Similar experiments with chabazite, which has a shorter unit cell than that of erionite, shows a similar periodicity (Figure 3.10) but the minimum *n*-paraffin diffusivity shifts down by about 2 carbon numbers, corresponding to the unit cell size difference between these two zeolites.

An example of this configurational diffusion effect on aromatics selectivity is the aforementioned achievement of para-directed aromatics alkylation and toluene disproportionation reactions (Chen et al., 1979). By increasing the crystallite size of ZSM-5, an increase in the diffusional constraints is imposed on the bulkier slower diffusing ortho and meta isomers, reducing the production of these isomers, and increasing the yield of the para isomer, as shown in Table 3.9.

A similar study on the effect of crystal size of HZSM-5 on the shape selectivity in xylene isomerization was made by Ratnasamy et al. (1986), starting with *m*-xylene. They also demonstrated that as the crystal size of HZSM-5

Table 3.9 Results with Large-Crystal ZSM-5

	Alkylation	Disproportionation	Thermodynamic equilibrium
Temp.,[a] °C	500	550	
WHSV[b]	6.6	30	
Feedstock	2:1 mol ratio of toluene/methanol	toluene	
Conversion, wt %			
toluene	39	13.2	
methanol	99		
Product distribution, wt %			
C_1–C_5	2.6	<0.1	
benzene	1.9	5.5	
toluene	54.0	86.8	
xylenes			
para	17.9	2.6	
meta	14.0	3.5	
ortho	7.0	1.4	
others	3.3	0.1	
Percent xylenes			
para	46	35	23
meta	36	46	51
ortho	18	19	26

[a]Weight hourly space velocity, (g of feed)/(g of catalyst) h^{-1}.
[b]Organic phase.
Source: Chen et al. (1979).

was increased from 8 to 16 μm, the para/ortho selectivity increased from 1.2 to 2.6. Kumar and Ratnasamy (1989) further showed that the shape selectivity in the reactions of xylenes increases from ZSM-5 to ZSM-48 to ZSM-22 and ZSM-23. It is interesting to note their data are consistent with the structure of zeolites shown in Table 2.1.

With the 10-ring ZSM-5 catalyst, Weisz et al. (1979) demonstrated that a triglyceride as large as $C_{57}H_{104}O_6$ can be converted to the same product spectrum of hydrocarbon products as methanol. This suggests that the triglyceride molecule can easily attain, through molecular dynamics, a critical dimension small enough to enter the channels of ZSM-5.

Benito et al. (1994) studied the dimerization of styrene to a mixture of dimers over large pore zeolites of Y and Beta. Among the dimers, including the linear and open 1,3-diphenyl-1-butenes and the cyclic 1-methyl-3-phenylindanes, they found no product selectivity with zeolite Y. However, with zeolite Beta, the trans isomer of 1,3-diphenyl-1-butene was selectively produced. Molecular graphic visualization of the isomers in the zeolite cages led them to conclude that the reaction over zeolite Beta was controlled by the diffusion of the organic molecules inside the cages.

IV. SPATIOSPECIFICITY, OR TRANSITION STATE SELECTIVITY

When both the reactant molecule and the product molecule are small enough to diffuse through the channels, but the reaction intermediates are larger than either the reactants or the products and are spatially constrained either by their size or by their orientation, we term this *spatiospecific selectivity* to distinguish it from stereo specificity (Haag et al., 1981). It is perhaps one of the most important properties of ZSM-5, making it superior to other catalysts in a number of important petrochemical processes. For example, Mobil's xylene isomerization process technology is built on ZSM-5's ability to suppress the disproportionation of xylene during the isomerization reaction.

Spatioselectivity, or transition state selectivity, is independent of crystal size and activity, but depends on the pore diameter and zeolite structure. This type of selectivity was first proposed by Csicsery (1971) when he observed the absence of symmetrical trialkylbenzenes in the product from the disproportionation of a dialkylbenzene over H-mordenite. Since the reaction is bimolecular, the diphenylmethane-type intermediates must require more space than is available in the mordenite channels. Alternative interpretations of the experimental results, on the basis of size exclusion and configurational diffusion inhibition, were ruled out by intramolecular isomerization experiments, which indicated that diffusion of symmetrical trialkylbenzenes in mordenite is not hindered. Product selectivity in the transalkylation and isomerization of methyl ethylalkylbenzene over a 12-oxygen ring opening mordenite was shown by Csicsery (1970, 1987) to be lacking the formation of symmetrical 1,3,5-isomers. The same reaction over faujasite produces these symmetrical isomers as the predominant products. The difference among zeolites was attributed to transition state selectivity.

The disproportionation of ethylbenzene to benzene and diethylbenzenes in 12-membered oxygen ring and 10-membered ring zeolites (Karge et al.,

Table 3.10   Distribution of Trimethylbenzene Isomers
Formed by Disproportionation of *m*-Xylene at 5.5%
Conversion

Zeolite	$X_{1,2,3}$	$X_{1,2,4}$	$X_{1,3,5}$	$\dfrac{X_{1,2,3}}{X_{1,3,5}}$
Y	7.9	64.0	28.1	0.26
Dealuminized Y	8.4	63.3	28.3	0.30

Source: Corma et al. (1988).

1984) showed that in 12-membered ring zeolites, the distribution of the diethylbenzenes is about 5% ortho, 62% meta, and 33% para isomer; on 10-ring zeolites, the ortho isomer is almost absent.

Similar transition state selectivity was used by Corma et al. (1988) to explain the change in para-to-ortho selectivity in the isomerization/disproportionation of *m*-xylene over dealuminized 12-membered oxygen ring faujasite. Unfortunately, the authors did not do the experiments on the effect of diffusivity on selectivity, because the small change in para-to-ortho ratio by dealumination of faujasite is consistent with the change in constraint index (as shown in Table 2.3) between dealuminized Y and rare earth exchanged undealuminized Y (see Table 3.10). One cannot rule out the configurational diffusion-controlled selectivity.

Spatiospecific selectivity plays a major role in the selective cracking of paraffins in medium pore zeolites. For example, *n*-hexane and 3-methylpentane are readily sorbed by ZSM-5, yet the singly branched molecule cracks at a significantly slower rate than the straight chain molecule (Chen and Garwood, 1978). 3-methylpentane, being bulkier than *n*-hexane, apparently requires more space than *n*-hexane to form the reaction intermediate, as shown in Figure 3.11. Definitive studies in support of spatial constraints of reaction intermediates rather than configurational diffusion inhibition of reactants were made by Frilette et al. (1981). They showed that the relative rate of cracking of *n*-hexane to that of 3-methylpentane is independent of the size of the individual catalyst crystal (Figure 3.12). Had diffusional constraint been an important factor in this reaction, the relative rate of cracking would have been affected by changing the crystallite size.

Amelse (1987), in studying the mechanism of transethylation over medium pore and large pore zeolites, concluded that the mechanism for transethylation is different because of the spatioselectivity of the medium pore

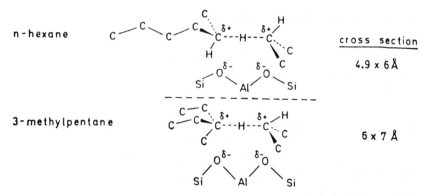

n-hexane

cross section

4.9 x 6 Å

3-methylpentane

6 x 7 Å

Figure 3.11 Mechanism of paraffin cracking. (From Haag et al., 1981.)

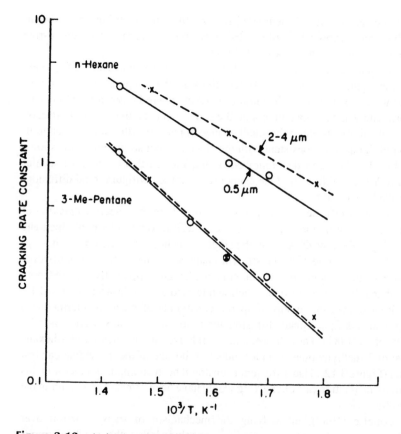

Figure 3.12 Arrhenius plot for the cracking of hexane isomers by HZSM-5 of different crystal size. (From Frilette et al., 1981.)

zeolites. The formation of a biphenylethane intermediate, which determines the rate of transethylation in large pore zeolites, is inhibited by the spatial constraints of the pores of the medium pore zeolites. Thus, the transethylation of ethylbenzene over medium pore zeolites takes place via a different mechanism, namely, dealkylation and realkylation of the ethyl group.

Kim et al. (1988) studied the alkylation of ethylbenzene with ethanol over ZSM-5 and concluded that the transition state selectivity plays the major role in forming the primary product, *p*-diethylbenzene. Similar experimental results were reported by Kaeding (1985), but he attributed his results to product selectivity due to configurational diffusion effect, because their para selectivities increased with increasing crystal size.

Weigert and Mitchell (1994) studied the isomerization of dimethylanilines over a variety of zeolites. They found that the pores in faujasite are large enough to allow isomerization of dimethylaniline to occur by *inter*molecular, hydrophilic, methyl-transfer reactions. Medium pore zeolites such as HZSM-5 only can equilibrate the three smallest dimethylanilines (2,4-, 2,5-, and 3,4-). It is interesting to note that the pores in zeolite H-Omega (ZSM-4), H-Beta and H-L are large enough to equilibrate all six dimethylanilines, albeit by an entirely different mechanism, namely *intra*molecular 1,2-methyl shifts.

V. TRAFFIC CONTROL

A number of zeolites, including ZSM-5, ferrierite, clinoptilolite, offretite, and mordenite, have intersecting channels of differing channel size (Table 2.1). Since the smaller channels are accessible only by the smaller molecules while the larger channels are accessible by both the large and small molecules, a new type of shape selectivity can be envisaged.

This concept was first proposed by Derouane, who termed it the *molecular traffic control effect* (Derouane and Gabelica, 1980). Derouane used this concept to explain the unusual absence of counterdiffusional effects in the ZSM-5 catalyzed conversion of simple molecules such as methanol. According to the concept of traffic control, the smaller molecules enter the sinusoidal channels, while the larger product molecules exit from the elliptical channels. The significance of this effect was tested by comparing the catalytic properties of ZSM-5 with ZSM-11, the latter having intersecting straight channels of the same size (Derouane et al., 1981). But for such reactions as the conversion of methanol to hydrocarbons and the alkylation of toluene with methanol, their results show no major difference in the catalytic activity and stability of these two

Table 3.11 Distribution of Cracked Products
from Catalytic Dewaxing

	TMA-offretite	ZSM-5
Methane	1	1
C_2	3	1
C_3	23	8
n-C_4	27	11
i-C_4	12	11
n-C_5	26	21
i-C_5	5	18
C_6	3	20
C_7^+	0	10

Source: Chen et al. (1984).

zeolites other than minor differences in product distribution, which thus largely negated the importance of this effect.

However, it is an interesting concept which should operate in any zeolite containing two intersecting channels of different channel sizes, since the smaller channels are accessible only by the smaller molecules while the larger channels are accessible by both the large and small molecules. For example, synthetic offretite, which has intersecting 12-ring and 8-ring channels, showed some interesting catalytic properties in the cracking of paraffins (Chen et al., 1984). We found it to be more selective to the cracking of n-paraffins than a 10-ring zeolite and also to yield more low molecular weight cracked products (Table 3.11).

We speculated that this is because of the blockage of the 12-ring channels by a small amount of stacking faults which reduce the accessibility of the larger branched paraffins and the availability of all the 8-ring channels to n-paraffins and other smaller molecules.

VI. CONFINEMENT EFFECT

The idea of confinement encompasses various interacting effects of the host on the guest molecules, including chemisorption, physisorption, and molecular orientation within the zeolite. This conformational relation between the reactant and the framework is proposed (Derouane et al., 1988a,b) to play an important role in the diffusion of the guest molecule in the zeolite and the reactivity of the guest molecule. An example of such an effect is the high selectivity for dehydrocyclization of hexane over the Pt-BaKL zeolite (Derouane and Vanderveken, 1988). By means of computational modeling of the van der Waals interaction between the zeolite framework and the guest molecule and molecular graphics, they attributed this high selectivity to the unique channel structure of zeolite L and the unique interaction of hexane with the platinum cluster. Only one platinum atom is accessible to hexane. That was taken to explain the extreme sensitivity of this catalyst to trace concentrations of sulfur. In the next step, the hexane molecule maximizes its van der Waals interaction with the channel walls and curls up in its most stable position. This is then followed by metal catalyzed C_6 ring closure and dehydrogenation to benzene. However, subsequent studies on the aromatization of *n*-hexane over Pt clusters supported on a high surface area MgO catalyst (Davis and Derouane, 1991) led to some doubt on the significance of the confinement effect on the selectivity of this reaction.

Several zeolites, such as ZSM-4 and mordenite, show remarkable activity and selectivity for the isomerization and disproportionation of light paraffins. For example, Guisnet et al. (1980) and Hilaireau et al. (1982) found that H-mordenite catalyzes *n*-butane disproportionation to produce equimolar yields of propane and pentanes. Chen (1986) reported the remarkable activity of ZSM-4 for these reactions, and showed that it is at least 20 times more active than mordenite. The only feature of these two zeolites that is different from other large pore zeolites is the shape of their channels. They are straight and do not contain either intersections or supercages. Whether this feature represents an example of the confinement effect that provides a specific site geometry promoting interaction of carbenium ions with reactant molecules remains to be proven.

REFERENCES

Amelse, J. A., "A Shape Selective Shift in the Mechanism of Transalkylation and Its Effect on the Ability to Hydrodeethylate Ethylbenzene," paper B-4, 10th North Am. Mtg. Catal. Soc., San Diego, May 17–22, 1987.

Benito, A., A. Corma, H. García, and J. Primo, Appl. Catal. A*116*, 127 (1994).

Chang, C. D., Catal. Rev.-Sci. Eng. *25*, 1 (1983); *Hydrocarbons from Methanol*, Marcel Dekker, New York, 1983.

Chen, N. Y., U.S. Pat. 3,732,326, May 8, 1973.

Chen, N. Y., J. Phys. Chem. *80*, 60 (1976).

Chen, N. Y., Proc. 7th Int. Zeol. Conf., Y. Murakami, A. Iijima, and J. W. Ward, eds., Kodansha/Elsevier, Tokyo/Amsterdam, p. 653 (1986).

Chen, N. Y. and W. E. Garwood, Advan. Chem. Ser. *121*, 575 (1973).

Chen, N. Y. and W. E. Garwood, J. Catal. *52*, 453 (1978).

Chen, N. Y. and J. N. Miale, U.S. Pat. 4,420,561, Dec. 13, 1983.

Chen, N. Y. and P. B. Weisz, Chem. Eng. Prog. Symp. Ser. *73*, 86 (1967).

Chen, N. Y., S. J. Lucki, and E. B. Mower, J. Catal. *13*, 329 (1969).

Chen, N. Y., W. W. Kaeding, and F. G. Dwyer, J. Am. Chem. Soc. *101*, 6783 (1979).

Chen, N. Y., J. L. Schlenker, W. E. Garwood, and G. T. Kokotailo, J. Catal. *86*, 24 (1984).

Corma, A., V. Fornés, J. Perez-Pariente, E. Sastre, J. A. Martens, and P. A. Jacobs, ACS Symp. Ser. *368*, 555, 1988.

Csicsery, S. M., J. Catal. *19*, 394 (1970).

Csicsery, S. M., J. Catal. *23*, 124 (1971).

Csicsery, S. M., J. Catal. *108*, 433 (1987).

Davis, R. J. and E. G. Derouane, J. Catal. *132*, 269 (1991).

Derouane, E. G. and Z. Gabelica, J. Catal. *65*, 486 (1980).

Derouane, E. G. and D. Vanderveken, Appl. Catal. *45*, L15 (1988).

Derouane, E. G., P. Dejaifve, and Z. Gabelica, Chem. Soc. Faraday Disc. *72*, 331 (1981).

Derouane, E. G., J. M. Andre, and A. A. Lucas, J. Catal. *110*, 58 (1988a).

Derouane, E. G., J. B. Nagy, C. Fernandez, Z. Gabelica, E. Laurent, and P. Maljean, Appl. Catal. *40*, L1 (1988b).

Frilette, V. J., W. O. Haag, and R. M. Lago, J. Catal. *67*, 218 (1981).

Gianneti, J. P. and A. J. Perrotta, Ind. Eng. Chem., Process Des. Develop. *14*, 86 (1975).

Goldstein, T. P., "The Silicon/Aluminum Ratio as a New Parameter in Determining Molecular Size Selective Sorption by Crystalline Aluminosilicates," paper presented at Am. Chem. Soc. 153rd Mtg., Miami Beach, FL, April 9–14, 1967.

Gorring, R. L., J. Catal. *31*, 13 (1973).

Guisnet, M., N. S. Gnep, C. Bearez, and F. Chevalier, Stud. Surf. Sci. Catal. *5*, 77 (1980).

Haag, W. O., R. M. Lago, and P. B. Weisz, Chem. Soc. Faraday Disc. *72*, 317 (1981).

Heck, R. H. and N. Y. Chen, Appl. Catal. A*86*, 83 (1992).

Hilaireau, P., C. Bearez, F. Chevalier, G. Perot, and M. Guisnet, Zeolites *2*, 69 (1982).

Jacobs, P. A., J. A. Martens, J. Weitkamp, and H. K. Beyer, Chem. Soc. Faraday Disc. *72*, 353 (1981).

Kaeding, W. W., J. Catal. *95*, 512 (1985).

Karge, H. G., Y. Wada, J. Weitkamp, S. Ernst, U. Girrbach, and H. K. Beyer, Stud. Surf. Sci. Catal. *19*, 101 (1984).

Kim, J.-H., S. Namba, and T. Yashima, Bull. Chem. Soc. Jpn. *61*, 1051 (1988).

Kumar, R. and P. Ratnasamy, J. Catal. *118*, 68 (1989).

Kumar, R., G. N. Rao, and P. Ratnasamy, Stud. Surf. Sci. Catal. *49B*, 1141 (1989).

Martens, J. A., J. Perez-Pariente, E. Sastre, A. Corma, and P. A. Jacobs, Appl. Catal. *45*, 85 (1988).

Miale, J. N., N. Y. Chen, and P. B. Weisz, J. Catal. *6*, 278 (1966).

Nakamoto, H. and H. Takahaski, Zeolites *2*, 67 (1982).

Nitsche, J. M. and J. Wei, AIChE J. *37*, 661 (1991).

Olson, D. H., W. O. Haag, and R. M. Lago, J. Catal. *61*, 390 (1980).

Olson, D. H., G. T. Kokotailo, S. L. Lawton, and W. H. Meier, J. Phys. Chem. *85*, 2238 (1981).

Ratnasamy, P., G. P. Rabu, A. J. Chandwadkar and S. B. Kulkarni, Zeolites *6*, 98 (1986).

Richter, M., H.-L. Zubowa, E. Schreier and J. Richter-Mendau, Zeolites *14*, 414 (1994).

Tatsumi, T., M. Nakamura, K. Yuasa, H. Tominaga, Catal. Lett. *10*, 259 (1991).

Weigert, F. J. and R. S. Mitchell, J. Mol. Catal. *89*, 210 (1994).

Weisz, P. B., Chemtech *3*, 498 (1973).

Weisz, P. B., V. J. Frilette, R. W. Maatman, and E. B. Mower, J. Catal. *1*, 307 (1962).

Weisz, P. B., W. O. Haag, and P. G. Rodewald, Science *206*, 57 (1979).

Weitkamp, J. and S. Ernst, Catal. Today *19*, 107 (1994).

Wu, E. L. and G. R. Landolt, "Pore Size and Shape Selective Effects in Zeolite Catalysis," paper presented at the 7th Int. Zeol. Conf., Tokyo, August 17–22, 1986.

4

Shape Selective Catalysis

I. REACTANTS OF INTEREST

It is generally accepted that hydrogen zeolites catalyze via carbenium ion intermediates, similar to reactions catalyzed by strong acids in homogeneous media. Past reviews have examined various reactions involving hydrocarbons, including paraffins, olefins, naphthenes, and aromatics and various nonhydrocarbons (Hölderich et al., 1988; Hölderich, 1991; Venuto, 1994; Dartt and Davis, 1995).

With 8-membered oxygen ring small pore zeolites, shape selective reactions deal only with the conversion of straight chain molecules. With the availability of medium pore zeolites, reactions involving non–straight chain molecules and nonhydrocarbon molecules are receiving increased attention.

In the following sections, we will examine the impact of shape selective catalysis on some of the more common organic reactions important to the petroleum and petrochemical industries. Among the reactants of interest are straight chain and slightly branched chain paraffins and olefins; benzene and alkylbenzenes; single ring naphthenes and alkyl naphthenes; and nonhydrocarbons, such as alcohols, ethers, acids, esters, pyridines, phenols, and others.

II. REACTIONS CATALYZED BY ZEOLITES

A. Olefin Reactions

Olefins undergo a variety of acid catalyzed reactions, including double bond isomerization, skeletal isomerization, oligomerization, transmutation or disproportionation, cracking/polymerization, hydrogen transfer, and cyclization or aromatization.

With medium pore zeolites, these reactions are taking place under the constraints imposed by the size of the channels. As a result, unique products are formed as a function of reaction temperature and pressure.

The application of these olefin reactions over ZSM-5 to the production of gasoline and distillate from light olefins led to the development of Mobil's Olefin to Gasoline and Distillate (MOGD) process (Tabak, 1984) and Olefin to Gasoline (MOG) process (Yurchak et al., 1990). Shell developed the Shell Poly-Gasoline and Kero (SPGK) process (van den Berg et al., 1989).

Skeletal isomerization of olefins, paraffins, and aromatics over aluminophosphate-based zeolites was reported by Pellet et al. (1988).

Isomerization

Olefins isomerize readily over acidic zeolite catalysts. For example, relative to the rate of cracking of *n*-hexane over HZSM-5, the rate of double bond shift of 1-hexene is at least 6 orders of magnitude faster (Haag et al., 1984), and is often diffusion controlled despite the fact that hexene is a relatively small molecule (Dessau, 1984). Consequently, the degree of branching of an olefin molecule will affect its rate of isomerization, and molecular shape selectivity among the olefinic molecules is to be expected.

Haag and Santiesteban (1993) found that Theta-1 (ZSM-22), ZSM-23, and ZSM-35 are excellent double bond isomerization catalysts. Catalyst aging is minimal despite operating at low temperatures and high pressures. Table 4.1 shows the aging data on the conversion of 1-butene to 2-butene. Simon et al. (1994) studied the skeletal isomerization of 1-C$_4^=$ over Theta-1 (ZSM-22) catalyst.

Isofin, a new process for olefin isomerization has been developed jointly by Mobil and BP (Kunchal et al., 1993).

Isomerization—double bond shift:

$$CH_3CH_2CH = CH_2 \ \rightleftharpoons \ CH_3 - CH = CH - CH_3$$

$$CH_2 = \underset{\underset{CH_3}{|}}{C} - CH_2 - CH_3 \ \rightleftharpoons \ CH_3 - \underset{\underset{CH_3}{|}}{C} = CH - CH_3$$

Table 4.1 Double Bond Isomerization of 1-Butene

Zeolite	WHSV	Temperature, °C	CSP
ZSM-5	18.6	120	<372
ZSM-22	9.9	125	>>356
ZSM-23	18.6	120	>>484
ZSM-35	37.4	122	>>5460

CSP = catalyst stability parameter = kg feed processed per kg catalyst
during useful cycle lift.
Source: Haag and Santiesteban (1993).

Skeletal isomerization:

$$CH_3-CH_2-CH_2-CH=CH_2 \rightleftharpoons CH_3-CH_2-\underset{\underset{CH_3}{|}}{C}=CH_2$$

Oligomerization

With medium pore zeolites, light olefins undergo rapid isomerization and oligomerization reactions at low temperatures with remarkable selectivity and stability not found with large pore zeolites.

$$2\,CH_3-CH_2-CH=CH_2$$
$$\rightleftharpoons CH_3-CH_2-\underset{\underset{CH_3}{|}}{CH}-CH=CH-CH_2-CH_3$$

For example, Garwood (1983) showed that propene reacts to form predominantly trimers, tetramers, and pentamers over HZSM-5 at 200°C, 2.7 WHSV, and 35 atm (Figure 4.1). At longer residence times, olefin transmutation reactions convert the oligomers to secondary olefinic products. Similar results at atmospheric pressures were reported by Haag (Figure 4.2) (Haag, 1984). The oligomerization activity of medium pore phosphate-based molecular sieves, such as SAPO-11 and SAPO-31, was reported by Long et al. (1985) and Pellet et al. (1987).

Although no detailed data are available on the isomer distribution of the oligomers, it is expected that the oligomers produced by medium pore molecular sieve catalysts are less branched than those produced by non–shaped selective acid catalysts (Haag et al., 1981). Van Hooff and co-workers (1983) showed that the oligomers formed in ZSM-5 at room temperatures from C_2 to

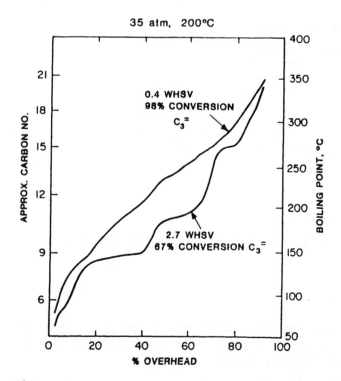

Figure 4.1 Effect of space velocity on propene conversion over HZSM-5. (From Garwood, 1983.)

C_4 olefins are linear in structure and proposed a mechanism involving stretching reactions via a cyclopropane intermediate to explain their formation (van den Berg et al., 1983). ^{13}C NMR analyses of propene oligomers also confirm qualitatively that the degree of branching is controlled by the dimensions of the zeolite cages and the pore structure (Galya et al., 1985).

Transmutation/Disproportionation

With medium pore zeolites such as HZSM-5, the secondary olefinic product from different feeds is substantially independent of the feed molecules. This suggests a rapid equilibration of olefins to a composition determined by the temperature and pressure of the system (Garwood, 1983).

The equilibration is generally perceived to be the result of a series of acid catalyzed polymerization and cracking reactions of the oligomers, involving intramolecular rearrangement of carbenium ions, dimerization, and β-scission.

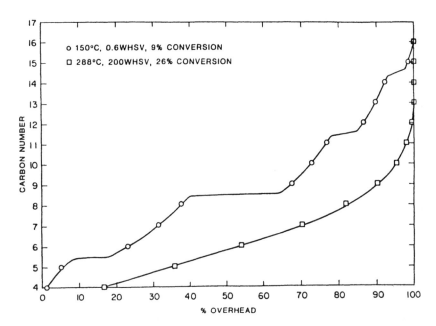

Figure 4.2 Effect of temperature and pressure on propene conversion. (From Haag, 1984.)

Transmutation:

$$C_4H_8 \rightleftharpoons \text{Mixture of } C_2\text{–}C_{10} \text{ olefins}$$

Disproportionation:

$$CH_3\text{—}CH_2\text{—}\overset{\overset{\textstyle CH_3}{\textstyle |}}{C}= CH\text{—}CH_2\text{—}CH_2\text{—}CH_3$$

$$\rightleftharpoons CH_3\text{—}CH_2\text{—}\overset{\overset{\textstyle CH_3}{\textstyle |}}{C}= CH_2 + CH_2 = CH\text{—}CH_3$$

For example, Garwood (1983) showed that at 275°C, propene, a mixture of pentenes, 1-hexene, and 1-decene all produced predominantly C_3–C_9 olefins (Table 4.2) over HZSM-5. They have a similar carbon number distribution and isomer distribution, which approaches chemical equilibrium. Table 4.3 compares the pentene product isomers with their equilibrium values.

Table 4.2 Olefin Equilibration Carbon Number Distribution

Product, wt %	Feed				
	Ethene	Propene	Pentene mix[b]	1-Hexene	1-Decene
Ethene	0[a]	<0.1	<0.1	<0.1	<0.1
Propene	11	8	10	9	4
Butenes	20	28	20	20	13
Pentenes	21	30	27	23	26
Hexenes	13	13	15	16	20
Heptenes	12	11	11	10	17
Octanes	8	6	7	8	8
Nonenes	8	3	5	6	7
$C_{10}^{'}$ (unidentified)	7	1	5	8	5

[a] Based on converted ethenes.
[b] 88% 2-methyl-2-butene, 8% 2-methyl-1-butene.
Source: Garwood (1983).

Table 4.3 Comparison of Percentage of Pentene Product Isomers with Equilibrium at 275°C

	Feed					
	Ethene	Propene	Pentene mix	1-Hexene	1-Decene	Equilibrium
1-Pentene	2	2	2	2	2	2
2-Methyl-1-butene	18	16	18	18	17	24
3-Methyl-1-butene	1	2	2	2	1	2
trans-2-pentene	11	10	11	12	13	9
cis-2-pentene	5	5	5	5	6	7
2-Methyl-2-butene	63	65	62	61	61	56

Source: Garwood (1983).

The hexene isomer distribution is shown in Figure 4.3 (Quann et al., 1986). When compared to the calculated equilibrium distribution, it is noted that at even as low as 3% conversion, the isomer distribution already approaches its equilibrium distribution. By following the extent of conversion, it can be seen that the reaction pathway begins with double bond shift followed by skeletal isomerization. The major deviation between the experimental and the calculated distributions are in the concentration of 2,3-dimethylbutenes, obviously the result of configurational diffusion constraints imposed by the zeolite.

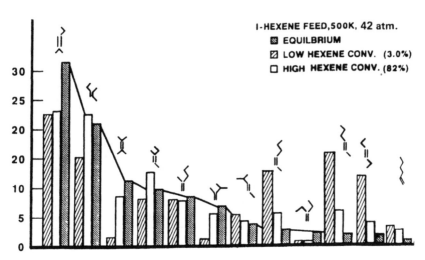

Figure 4.3 Comparison of experimental hexene isomer distribution with calculated equilibrium distribution. (From Quann et al., 1986.)

Tabak et al. (1984, 1986) attempted to calculate the theoretical equilibrium composition in the range of temperature and pressure of interest. To avoid handling an astronomical number of isomers involved, the calculation was simplified by lumping each group of isomers of the same carbon number as a single compound and calculating the equilibrium among the lumped groups. To obtain the lumped free energy of a group requires the knowledge of the free energy of all the individual isomers, but they are at present available only up to six carbon atoms. Thus, for the higher molecular weight groups, an approximation technique developed by Alberty (1983) was used.

Because not all the isomers that theoretically exist can be produced by the ZSM-5, it was necessary to modify the parameters used in the extrapolation of free energies to obtain a reasonable agreement with the experimental results. Table 4.4 compares the theoretical values with the experimental data shown in Table 4.2, obtained at 275°C. The agreement is reasonably good.

At higher pressures, the product carbon number and boiling ranges increase with reaction temperature (Figure 4.4). For example, an olefinic mixture, consisting of 5 wt % butenes, 30 wt % pentenes, and 47 wt % hexenes (remaining 18% consisting of 12% C_4–C_6 paraffins, 6 % C_7^+) was passed over HZSM-5 at 47 atm. As the reaction temperature was increased to 250°C, olefins boiling above 200°C became a significant part of the reaction product. Structural analysis of the 200°C$^+$ products shows primarily methyl branching with one methyl branch for each five carbons.

Table 4.4 Comparison of Calculated and Observed Olefin Equilibria

Charge products, wt %	Propene		Pentene		Hexene	
	Data	Calc.	Data	Calc.	Data	Calc.
Ethene	<0.1	0.7	<0.1	0.5	<0.1	0.4
Propene	8	5.3	10	4.1	9	3.8
Butenes	28	23.9	20	20.2	20	19.1
Pentenes	30	21.2	27	19.6	23	19.1
Hexenes	13	23.2	15	23.6	16	23.6
Heptenes	11	12.5	11	13.9	10	14.3
Octanes	6	6.5	7	8.0	8	8.4
Nonenes	3	3.4	5	4.5	6	4.9
C_{10}^+ (unidentified)	1	3.5	5	5.6	8	6.4

Source: Tabak et al. (1986).

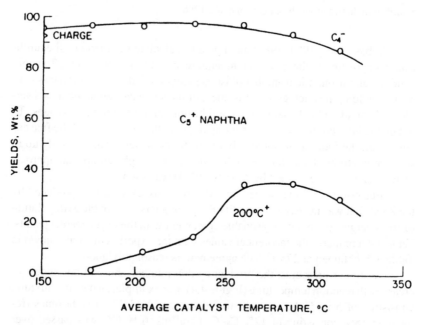

Figure 4.4 High boiling products from C_4, C_5, C_6 olefin mixture 1 LHSV, 47 atm. (From Garwood, 1983.)

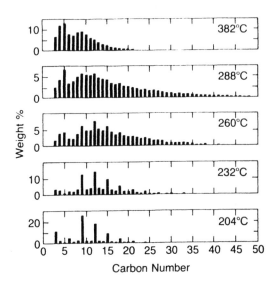

Figure 4.5 Propene polymerization at 55 atm, 1 WHSV. (From Quann et al., 1986.)

At higher temperatures the molecular weight distribution of the product decreases. This is shown in Figure 4.5 (Quann, 1986) for propene at 55 atm and 1 WHSV over the temperature range of 204 to 382°C.

Results of the theoretical equilibrium calculations made at elevated pressures, when compared with the experimental results as shown in Figure 4.6, show a large discrepancy at low temperatures. The calculation overestimates the average carbon number of the product at below 330°C. Nevertheless, it shows the trend of decreasing average carbon number at higher temperatures. The calculated product distribution agrees reasonably with the experimental data at above 330°C.

Hydrogen Transfer/Cyclization

In addition to transmutation/disproportionation reaction, olefins undergo hydrogen transfer reactions to form cyclo-olefin. The formation of cyclo-olefins (C_nH_{2n-2}) from the reaction of 1-hexene over ZSM-5 is shown in Figure 4.7.

$$2 \text{ C—C—C—C}=\text{C—C} \rightleftharpoons \bigcirc + \text{RH}$$

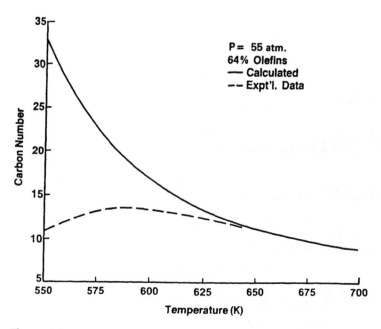

Figure 4.6 Average carbon number at equilibrium. (From Tabak et al., 1986.)

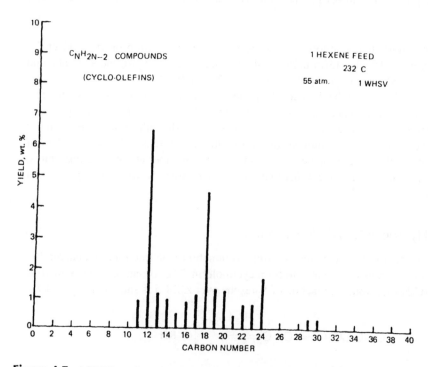

Figure 4.7 MOGD product distribution by FIMS. (From Quann et al., 1986.)

The presence of cyclo-olefins was verified by hydrogenation of selected products and subsequent examination by FIMS (Quann, 1986).

Hydrogen Transfer/Aromatization—M2-Forming

When the temperature is increased, a shift in product distribution toward BTX aromatics and light paraffins occurs (Table 4.5) as hydrogen transfer and cracking reactions become appreciable (Garwood, 1983).

Aromatization reactions of olefins become significant at about 370°C at atmospheric pressures over HZSM-5. M2-Forming, a generic name, has been coined to describe the reactions producing BTX aromatics from nonaromatic hydrocarbons over the medium pore zeolites (Chen and Yan, 1986).

The reaction products are again found to be independent of feed composition and are formed in a consecutive reaction pathway via cracking and hydrogen transfer reactions. The yield of aromatics is subject to the stoichiometric constraint of the carbon/hydrogen balance.

$$C_3{=} \; \rightleftharpoons C_2\text{--}C_{10} \; \text{olefins} \xrightarrow{\begin{array}{c} \text{Hydrogen} \\ \text{transfer} \end{array}} B{\cdot}T{\cdot}X + C_2H_6 + CH_4 + H_2$$

Table 4.5 Conversion of Propene at 390°C, 1 atm, 0.5 WHSV

Carbon no.	Product, wt %				
	Olefin	Isoparaffin	*n*-Paraffin	Aromatics	Total
1	—	—	0.2	—	0.2
2	—	—	0.8	—	0.8
3	0.4	—	23.3	—	23.7
4	0.7	17.1	9.1	—	26.9
5	0.1	7.2	1.7	—	9.0
6	0.1	1.0	0.2	2.0	3.3
7	0.1	0.9	0.2	10.7	11.9
8	—	—	—	12.8	12.8
9	—	—	—	6.3	6.3
10	—	—	—	2.1	2.1
11$^+$	—	—	—	3.0	3.0
Totals	1.4	26.2	35.5	36.9	100.0

Source: Garwood (1983).

It was observed that the channel intersections provide a large spherical space for two C_3–C_5 straight chain molecules to stack on top of each other and proposed a rationale for the formation of aromatics from aliphatics and the observed cutoff point of the aromatics that are made with the medium pore zeolites (Derouane and Vedrine, 1980; Dejaifve et al., 1980).

Ester Formation by Addition to Carboxylic Acids

The addition of an olefin to a carboxylic acid over acidic catalysts generally yields a mixture of esters. Young (1982) showed that α-methylalkyl carboxylates can be selectively produced over medium pore zeolites from carboxylic acids with either a terminal or an internal double bond.

$$R—CH=CH—CH_3 \rightleftharpoons R=CH_2 \rightleftharpoons R—CH=CH—R$$

$$+ R—C \begin{array}{c} O \\ OH \end{array}$$

$$R—C \begin{array}{c} O \\ O \end{array}$$

$$CH_3—CH—R$$

Table 4.6 shows the results of adding 1-octene to acetic acid in a batch autoclave at 16–20 atm. It is interesting to note that because of the fast double bond shift reaction and the size difference between the smaller α-methyl heptylacetate (2-octylacetate) and the other bulkier isomers (their relative rate of diffusion in medium pore zeolites being, therefore, very different), 2-octylacetate, a product of 1-octene and acetic acid, was selectively produced over ZSM-5 and ZSM-12. The selective production of α-methyl heptylacetate was even more dramatically demonstrated by using a mixture of octenes consisting of 25% 1-octene, 25% *trans*-2-octene, 25% *trans*-3-octene, and 25% *trans*-4-octene. As shown by the data in Table 4.7, instead of getting an isomeric mixture of octylacetates, 2-octylacetate was selectively produced over ZSM-12.

In a separate study, Sato (1984) showed that when compared to zeolite Y and mordenite, HZSM-5 is a far better catalyst for the production of ethylacetate and isopropylacetate—by adding to acetic acid ethene and propene, respectively:

$$+ \ C_2H_4 \ \rightleftharpoons \ CH_3-C\overset{\displaystyle O}{\underset{\displaystyle O\ C_2H_5}{{\Big\langle}}}$$

$$CH_3-\overset{\displaystyle O}{\underset{\displaystyle OH}{{\Big\langle}}}$$

$$+ \ C_3H_6 \ \rightleftharpoons \ CH_3-\overset{\displaystyle O}{\underset{\displaystyle O-iC_3H_7}{{\Big\langle}}}$$

Hydration

Chang and Morgan (1980) showed that hydration of C_2 to C_4 olefins to alcohols can be carried out over ZSM-5 at below about 240°C and 10 to 20 atmospheres of pressure without forming ethers or other hydrocarbons.

$$C_3H_6 + H_2O \longrightarrow iC_3H_7OH$$

Table 4.6   Reaction of 1-Octene with Acetic Acid

Catalyst: ZSM-12, $SiO_2/Al_2O_3 = 70$
Acetic acid/1-octene—4/1 molar

Reaction time, h	Temperature, °C	Pressure, atm	Yield, wt %	$C_{18}H_{17}OAc$ isomer, wt %		
				2	3	4
3.5	150	12	5.9	96.9	2.9	0.2
5.6	200	17	23.9	94.1	5.5	0.4
73.2	200	17	32.2	91.4	8.1	0.5

Source: Young (1982).

Table 4.7   Reaction of Mixed Octenes with Acetic Acid

Feed: 25% 1-octene, 25% *trans*-2-octene,
25% *trans*-3-octene and 25% *trans*-4-octene

Reaction time, h	Temperature, °C	Pressure, atm	Yield, wt %	$C_{18}H_{17}OAc$ isomer, wt %		
				2	3	4
1.75	150	12	14.1	88	11	1
2.75	150	12	17.0	86	13	1

Source: Young (1982).

Above 240°C, however, propene and butenes undergo other olefinic reactions forming higher molecular weight hydrocarbon products. Chang and Hellring (1986) reported that glycols can be selectively synthesized by the hydrolysis of olefin oxides at 25°C over medium pore zeolites with excellent yields.

Hydration of propene to diisopropyl ether (DIPE) using zeolite Beta as the catalyst was studied by Huang et al. (1990). MCM-22 was found to be an excellent catalyst to co-produce alcohol and ether from hydration of light olefins (Bell et al., 1990).

B. Paraffin Reactions

Cracking and Hydrocracking

The selective cracking of straight chain paraffins from a mixture of other hydrocarbons was one of the initial demonstrations of shape selective catalysis. The first commercial shape selective catalytic process, the Selectoforming process, uses erionite as the catalyst to selectively crack low octane n-paraffins in reformates (Chen et al., 1968).

Erionite is an effective cracking catalyst for gasoline range paraffin molecules (C_{11} and lower) under moderate pressures, but requires much higher hydrogen pressures (above 68 atm) to crack higher boiling paraffins. For example, at 35 atm, there is a sharp break in the relative rate of conversion of n-paraffins to non-normals (Chen and Garwood, 1978a). This pressure dependency is probably related to the diffusional characteristics of erionite (Gorring, 1973; Chen et al., 1968) but is not well understood.

Unlike erionite, the rate of cracking of paraffins over medium pore zeolites increases with the length of the molecules and decreases with the bulkiness of the molecules. Figure 4.8 shows the relative rate of cracking of each component when a mixture of C_5 to C_7 paraffins was passed over HZSM-5 at 1.4 LHSV, 35 atm, and 340°C (Chen and Garwood, 1978b).

As mentioned in Chapter 3, *spatiospecific* selectivity and configurational diffusion effects influence dramatically the selective cracking of paraffins in the ZSM-5 family of zeolites. The former effect is particularly important in the relative rate of cracking of normal and mono-methyl substituted paraffins, while the latter plays a major role in the cracking of multiple branched paraffins.

The extensive study on the mechanism of hydrocracking of long chain paraffins, by Jacobs, Weitkamp, and their co-workers (Jacobs et al., 1981; Martens, 1985; Weitkamp et al., 1983), showed that in hydrocracking over Pt/Ultrastable Y, the relative rates of all six different types of carbenium ion re-

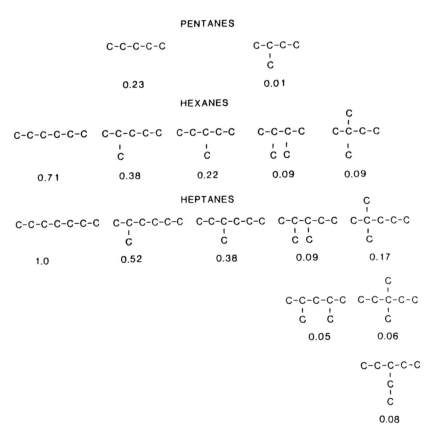

Figure 4.8 Relative rate of cracking of paraffins over ZSM-5 1.4 LHSV, 35 atm, 340°C. (From Chen and Garwood, 1978b.)

actions (isomerization of type A and B and the four types of hydrocracking A, B1, B2, and C) are independent of chain length above C_{10}.

Hydrocracking of type A is the fastest reaction, followed in decreasing order by isomerization of type A, B1-cracking, B2-cracking, B-type isomerization, and C-cracking, except that with increasing chain length, cracking of tribranched molecules via B1, B2, and C types is slightly faster than A-type, because the latter requires an α,α,γ-configuration which needs several alkyl shifts to form. This explains the observed increased selectivity for dibranched cracked products in heavier fractions.

The primary cracked product of the four different types of β-scission reactions denoted with A, B1, B2, and C can be distinguished as they require specific configurations of the side chains. In the case of octane, nonane, and decane, 35%, 53%, and 69% of the cracked products respectively are formed via the A type. The remaining products are formed via B1-, B2-, and C-cracking.

Detailed analyses of the composition of cracked products from the hydrocracking and hydroisomerization of n-paraffins over ZSM-5 (Weitkamp et al., 1983) suggest that reactions involving tertiary carbenium ion intermediates do not take place in medium pore zeolites. The relative high yield of C_3s, typical of medium pore and small zeolites is the consequence of β-scission of secondary carbenium ions, or to a lesser extent the not favored primary carbenium ions.

The shape selective conversion of paraffins is the basis for a number of commercial petroleum refining processes, including M-Forming, a second generation postreforming process (Heinemann, 1977) for upgrading the octane rating of reformates; MDDW, a distillate dewaxing process for increasing the quality and the yield of jet fuels, kerosene, and other distillate fuels (Chen et al., 1977; Donnelly and Green, 1980; Chen and Garwood, 1986); and MLDW, a catalytic lube dewaxing process, which replaces solvent dewaxing (Smith et al., 1980).

In addition to undergoing secondary cracking reactions, the olefinic cracked fragments also oligomerize, transmutate, cyclize, aromatize, and hydrogen transfer, forming naphthenes, aromatics, and lower molecular weight paraffins. Alkylation of the aromatics and a minor amount of ring closure to form dicyclics are also indicated. These can be seen in Table 4.8, which lists these products for n-octane conversion over ZSM-5 at 35 atm and 275°C (Chen et al., 1979a).

Note also that the n-octane feed and its products are in almost quantitative hydrogen balance (Table 4.9). This balance explains the almost complete absence of coke on the catalyst and the remarkable stability of the catalyst in the absence of a hydrogenation component and gaseous hydrogen. This is unique to the medium pore zeolites.

Isomerization and Hydroisomerization

Hydroisomerization of light paraffins, such as C_4 to C_8 paraffins over medium pore zeolites, has been patented by Mobil (Haag and Lago, 1983a). With enhancement of activity by steam and/or ammonia (Haag and Lago, 1983b), Pt/ZSM-5 was shown to be an excellent catalyst for the isomerization of n-

Table 4.8 Products from *n*-Octane Cracking 275°C, 35 atm, 1 LHSV

	wt %
Methane	< 0.1
Ethane	0.1
Propane	4.1
Propene	0.1
i-Butane	4.9
n-Butane	7.9
Butenes	0.1
i-Pentane	5.8
n-Pentane	9.6
C_6 Paraffins	4.8
C_7 Paraffins	2.4
C_8 Paraffins	49.7
Mono olefins	0.9
Diolefins	0.9
Benzene	1.1
Toluene	0.2
C_8 Alkylbenzenes	1.1
C_9 Alkylbenzenes	1.2
C_{10} Alkylbenzenes	0.6
C_{11} Alkylbenzenes	0.5
C_{12} Alkylbenzenes	0.4
C_{13} Alkylbenzenes	0.3
Tetralins, indanes	0.6
Naphthalenes	0.2
Monocyclic naphthenes	2.3
Bicyclic naphthenes	0.1
Mono olefins	0.9
Diolefins	0.9
Total	100.0

Source: Chen et al. (1979a).

pentane. The stability of the zeolite against coking allows the reactions to be carried out at lower hydrogen-to-hydrocarbon ratios and lower pressures.

Hydroisomerization of long chain paraffins, including C_9 C_{16} paraffins, was extensively studied by Jacobs, Weitkamp, and their co-workers (Jacobs

Table 4.9 Hydrogen Balance

	n-Octane charge, 100 g	Products
Hydrogen content, g	15.89	15.85

Source: Chen et al. (1979a).

et al., 1981; Weitkamp et al., 1983). Based on detailed product distributions obtained from C_7–C_{16} paraffins, the following reaction network was proposed:

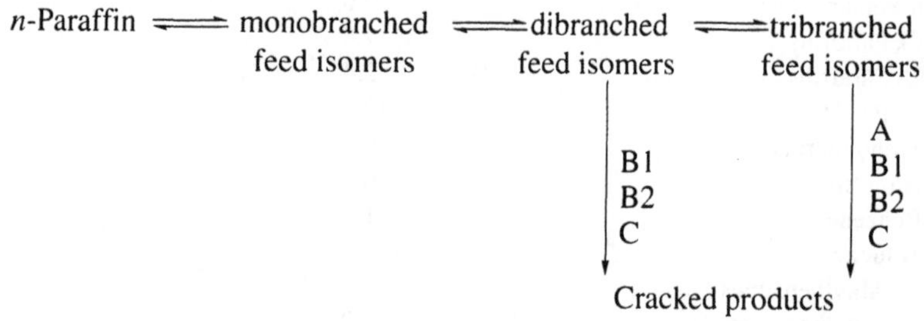

n-Paraffin ⇌ monobranched ⇌ dibranched ⇌ tribranched
feed isomers   feed isomers   feed isomers

                                          A

                     B1   B1

                     B2   B2

                      C   C

Cracked products

In this network the isomerization reaction proceeds via two routes: (1) type A isomerization (via alkyl shift), which is faster than (2) type B isomerization (via protonated cyclopropane). A, B1, B2, and C are four different types of β-scission reactions shown in Table 4.10.

Starting with normal and monobranched paraffins, isomerization clearly precedes hydrocracking reactions. Cracking of normal as well as of monobranched alkylcarbenium ions is a very slow process. Cracking starts only when the n-paraffin molecule has undergone at least two isomerization steps. Figure 4.9 shows that over a large pore Pt/Ultrastable Y catalyst (Martens, 1985), high yields of isomerized product can be obtained before cracking becomes a significant factor. However, as the isomer assumes an α,α,γ-configuration, it is hydrocracked rapidly via the type A cracking route (tertiary to tertiary).

Compared to large pore zeolites, such as the Ultrastable Y (Jacobs et al., 1980), the channel structure of ZSM-5 was found to exert a pronounced effect on both the relative rate of cracking to isomerization and on the isomer distribution.

At 1 atm, with the Ultrastable Y catalyst, decane isomerizes without cracking; with medium pore zeolites, the formation of bulky di- and tribranched decanes and ethyloctanes is suppressed. For example, at 5% conversion of

Table 4.10 Possible β-Scission Mechanisms on Secondary and Tertiary Carbocations

Type	Ions involved	Example
A	tert[a] → tert	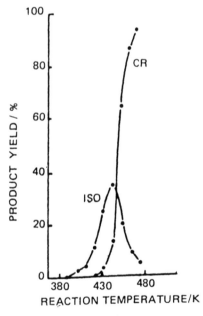
B1	sec[b] → tert	
B2	tert → sec	
C	sec → sec	

[a]tert = tertiary.
[b]sec = secondary.
Source: Weitkamp et al. (1983).

Figure 4.9 Yield of isomers (ISO) and cracked products (CR) from decane, converted over Pt/USY at various reaction temperatures. (From Martens, 1985.)

decane, with Pt/ZSM-5 ethyloctanes are not produced and less than 10% of the isomers are dibranched. By increasing the pressure to 20 atm. (Weitkamp et al., 1983), Pt/ZSM-5 can isomerize *n*-nonane without any hydrocracking; however, *n*-pentadecane still cracks extensively and gives low yields of skeletal isomers, which consist primarily of mono-methyl paraffins.

With more recent studies on acid/metal balanced medium pore zeolites such as Pt/ZSM-5 (Chen and Walsh, 1989a), Pt/ZSM-11 (Chen and Walsh, 1989b), and Pt/SAPO-11 (Miller, 1989, 1993) catalysts, the hydroisomerization selectivity can be significantly improved over the same catalysts without the metal function as indicated by the greater yield of the liquid product at the same pour point. Results for the bifunctional catalyst compared to the monofunctional dewaxing catalyst are shown in Figures 4.10 and 4.11.

The ability of Pt/Zeolite Beta to hydroisomerize high molecular weight paraffins in a mixture with aromatics (LaPierre et al., 1983, 1985) made it possible for Mobil to develop an isomerization process known as the Mobil Isomerization Dewaxing (MIDW) process (Buyan et al., 1995) to lower the pour point of gas oils. Table 4.11 shows a comparison of gas oil dewaxing over a va-

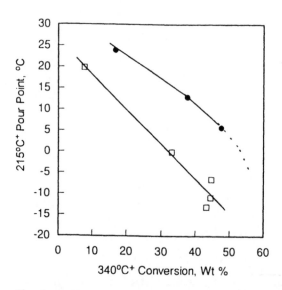

Figure 4.10 Comparison of 215°C⁺ pour point versus 340°C⁺ conversion for Pt/ZSM-5, (□) and Ni/ZSM-5 (●). Experimental conditions: pressure = 28, atm LHSV = 1, feedstock: hydrotreated 30°C pour point heavy coker gas oil. (From Chen and Walsh, 1989a.)

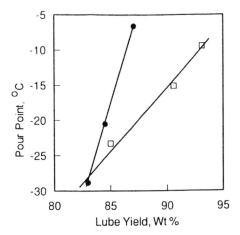

Figure 4.11 Comparison of pour point versus lube yield for Pt/SAPO-11 (□) and ZSM-5 (●). Experimental conditions: pressure = 150 atm, LHSV = 1, feedstock: 24°C pour point lube oil. (From Miller, 1989.)

riety of zeolite catalysts, including zeolite Beta, ZSM-20 and mordenite. Zeolite Beta gave higher liquid yield and lower pour point than the other catalysts and demonstrated its high selectivity for isomerization.

Aromatization—M2-Forming

At higher temperatures (up to 550°C) and atmospheric pressures, M2-Forming of paraffins to form predominantly BTX aromatics with hydrogen, methane, and ethane as the major byproducts proceeds readily over medium pore zeolites.

$$C_nH_{2n+2} \longrightarrow C_mH_{2m+2} + C_bH_{2b}$$

$$[C_2\text{---}C_{10} \text{ olefins}]$$

$$(CH_2)_xC_6H_6 \quad (BTX)$$
$$+$$
$$CH_3 + C_2H_6 + H_2$$

Table 4.11 Hydroisomerization of Hydrotreated Light Gas Oil (215–380°C)

	Pt/Beta	Pt/ZSM-20	Pd/mordenite
Catalyst			
Pressure, atm	35	52	35
Temperature, °C	315	370	315
LHSV	1	1	0.5
Products, %			
C_4^-	1.8	4.6	6.8
C_5^-165°C	16.5	24.8	53.3
165°C$^+$	81.7	70.6	39.9
Total liquid product			
Pour point, °C	−65	−39	−42

Source: LaPierre et al. (1985).

The yield of aromatics increases with decreasing hydrogen content of the feed, again subject to the stoichiometric constraints of the carbon/hydrogen balance (Chen and Yan, 1986).

Dehydrogenation/Aromatization with Dual Functional Catalysts

The introduction of a dehydrogenation function to the zeolites increased significantly the conversion of aromatics from light paraffins by the increase in the dehydrogenation activity (Kwak et al., 1994). The stability of these metals, zinc and gallium, and their oxidation states are the subject of considerable interest. Plank et al. (1979) studied the stabilization of zinc with other metals for the aromatization reactions. Kitagawa et al. (1986), Price and Kanazirev (1990), Abdul Hamid et al. (1994), Giannetto et al. (1994), and Kwak et al. (1994) studied the role of gallium and platinum in zeolites in the conversion of light paraffins to aromatics. The improvement in aromatics selectivity compared to HZSM-5 is shown in Table 4.12. Kwak et al. stated that under low hydrogen pressure platinum is rapidly deactivated by the added or *in situ* generated 5-membered ring naphthenes, while Ga_2O_3 and ZnO containing HZSM-5 catalysts were not deactivated (Figure 4.12)

Chang et al. (1991) studied the incorporation of gallium to large pore zeolites, such as zeolite Beta, and demonstrated the increase in aromatization activity.

Table 4.12 Aromatics Selectivity at
Various Propane Conversion at a Constant
Temperature of 530°C

Sample	Aromatics selectivity (mol %) at propane conversion of		
	10%	20%	40%
H	0.5	1.5	5
Ga	17	20	22
Pt	3	4	6

Source: Kwak et al. (1994).

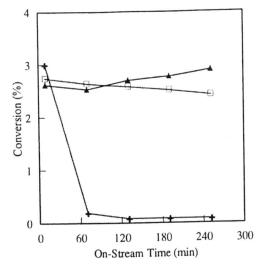

Figure 4.12 Time-on-stream dependence of propane conversion: (▲) H, (□) Ga, (+) Pt. (From Kwak et al., 1994.)

C. Reactions of Aromatic Compounds

Aromatic compounds including alkylbenzenes and polycyclics undergo a variety of reactions over acid catalysts including isomerization, transalkylation/ disproportionation, alkylation/dealkylation, and paring reactions (Sullivan et al., 1961). It is expected that the pore structure of the zeolites would have a major influence on their stability and product distribution.

Isomerization

HZSM-5

HZSM-5 shows an outstanding ability to isomerize xylenes with a minimum of side reactions, such as disproportionation to toluene and trimethylbenzenes. Haag and co-workers (Haag and Dwyer, 1979; Olson and Haag, 1984) have shown that the observed high isomerization selectivity is probably the result of transition state selectivity, i.e., the bimolecular disproportionation of xylenes, which involves a bulky diphenylmethane type reaction interme diate that is sterically difficult to accommodate in the pores of the zeolite, and is hindered relative to the monomolecular isomerization reaction (see Figure 4.13).

Based on these findings, several xylene isomerization processes have since been developed by Mobil.

It is also interesting to note that because of the large difference in the diffusion coefficient of p-xylene and o- and m-xylenes, $D_p/D_{o,m} > 10^4$, diffusional effects can give isomerized products containing p-xylene in concentrations exceeding that of its equilibrium value. An example of such diffusion-disguised kinetics is shown in Figure 4.14. In this case, the reaction pathway of o-xylene isomerization over the ZSM-5 catalyst was changed by treating the catalyst chemically to reduce its diffusion coefficient. It is noted that in the case of the chemically treated catalysts, the ratio of *para-* to *meta*-xylene exceeds its equilibrium value (Young et al., 1982). Similarly, high *para-* to *ortho*-xylene was observed when xylene was isomerized over a large crystal ZSM-5 (Ratnasamy et al., 1986).

Ratnasamy et al. (1989) studied the isomerization of m-xylene over the larger pore zeolite Beta. No product shape selectivity was observed. However shape selectivity was observed in the methylation of toluene and in the disproportionation of m-xylene and toluene. Zeolite Beta was able to discriminate among the trimethylbenzenes and suppressed the formation of the bulkier 1,2,3 and 1,3,5 isomers.

Platinum containing gallium-substituted ZSM-12, on the other hand, was claimed by Sachtler et al. (1992) to be a xylene isomerization catalyst.

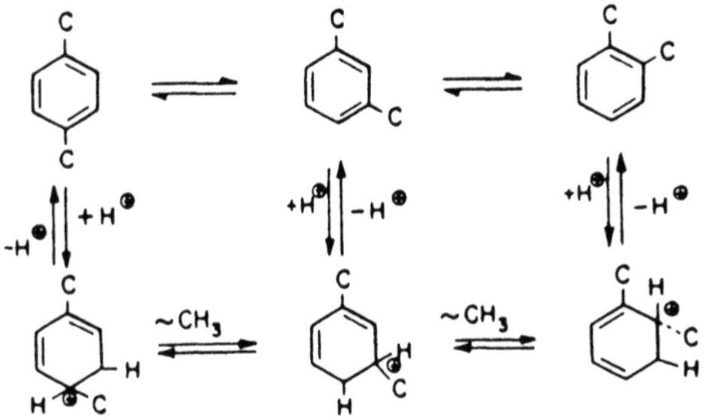

(a)

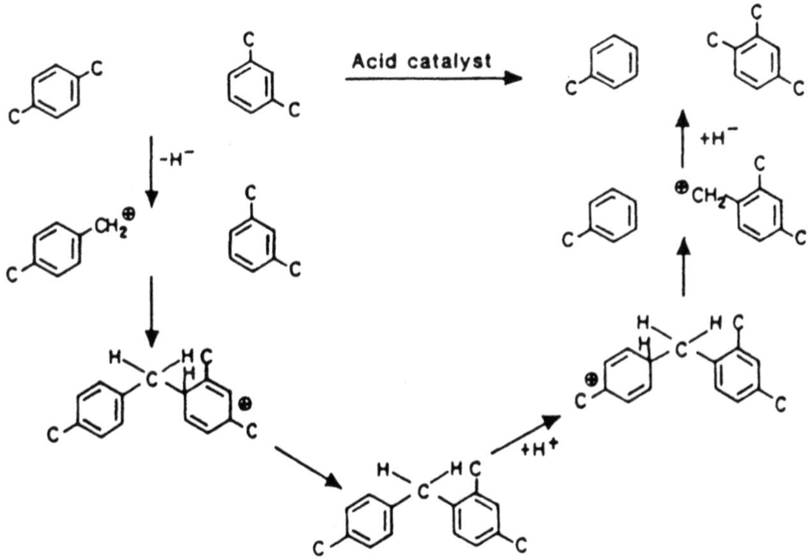

(b)

Figure 4.13 Comparison of monomolecular and biomolecular reactions of xylenes. (a) Acid-catalyzed xylene isomerization mechanism. (b) Acid-catalyzed xylene disproportionation mechanism. (From Olson and Haag, 1984.)

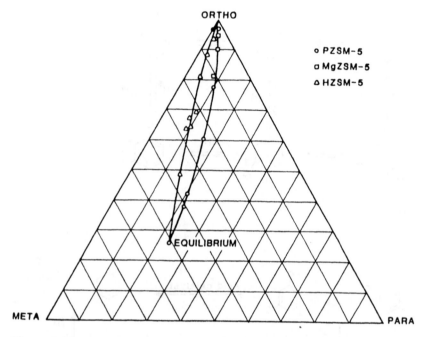

Figure 4.14   Reaction paths for isomerization of *o*-xylene. (From Young et al., 1982.)

H-mordenite

Song and Moffatt recently (1993) studied the ring-shift isomerization of oc-tahydrophenanthrene (sym-OHP) to octahydroanthracene (sym-OHAn) over larger pore (12-membered ring) zeolites, Y, and mordenite. They found the H-mordenite selectivity catalyzed the ring-shift reaction.

sym—OHP   sym—OHA

Transalkylation/Disproportionation

Because ethylbenzene is always present in commercial xylene streams to the xylene isomerization process, the relative rate of transalkylation of ethylbenzene to that of disproportionation of xylene is an important factor in determining the selectivity of the catalyst.

The transalkylation of ethylbenzene proceeds readily over the medium pore zeolites. Karge and co-workers (Karge et al., 1982, 1984) used ethylbenzene transalkylation as a test reaction for the determination of acid sites in ZSM-5 and ZSM-11 and found a linear relationship between conversion and the aluminum content of the zeolite. As can be expected, ZSM-5 and ZSM-11 produced high concentrations of *p*-di-ethylbenzene and no *o*-di-ethylbenzene (Kaeding, 1985). This is unlike mordenite, which gave an essentially equilibrated di-ethylbenzene composition.

Transalkylation of C$_8$ aromatics takes place between two ethylbenzene molecules and between an ethylbenzene and a xylene molecule to produce a mixture of benzene, toluene, trimethyl, methyl-ethyl, diethyl, and dimethyl-ethylbenzenes, Figure 4.15.

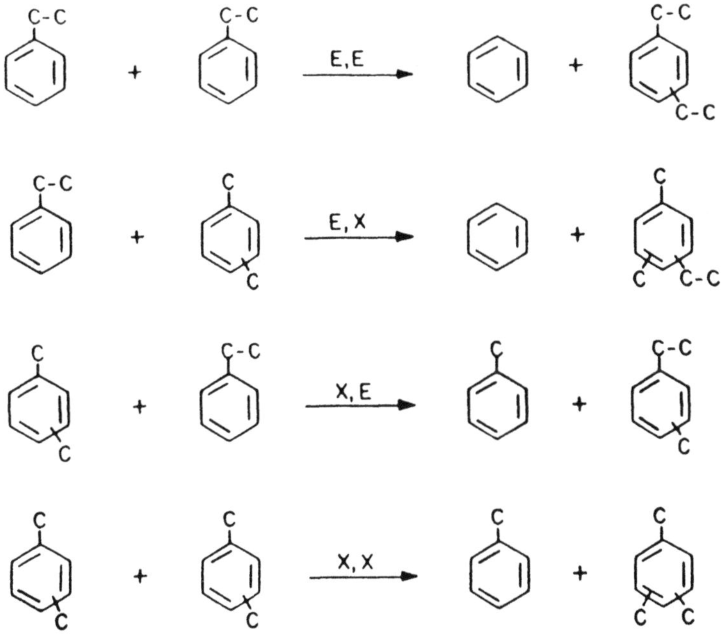

Figure 4.15 Transalkylation reactions in the ethylbenzene–xylene system. (From Olson and Haag, 1984.)

Table 4.13 Selectivity in Xylene Isomerization

Feed: 15% Ethylbenzene
      85% Xylene (63% *meta*, 22% *ortho*)

Catalyst	% Xylene transalkylated / % Ethylbenzene transalkylated
ZSM-4	0.36
Mordenite	0.19
ZSM-5	0.09

Source: Olson and Haag (1984).

From the product composition, the relative rates of these reactions over ZSM-5, mordenite, and ZSM-4 have been determined (Olson and Haag, 1984). As shown in Table 4.13, the data clearly indicate that the selectivity of ZSM-5, mordenite and ZSM-4 is quite different for these transalkylation/disproportionation reactions. These results demonstrate the pronounced effect of zeolite pore structure on reaction selectivity. Transalkylation of C_8 aromatics in ZSM-5 occurs predominantly between ethylbenzene molecules, rather than between ethylbenzene and xylenes, the latter reaction occurs readily in larger pore zeolite catalysts. The more selective ethylbenzene transalkylation in ZSM-5 is a result of the faster diffusion rate of ethylbenzene relative to that of *m*- or *o*-xylene.

Toluene and $C_9{}^+$ aromatics undergo transalkylation reactions over ZSM-5 (Brennan and Morrison, 1976) to produce a mixture of xylenes.

Table 4.14 shows the result of transalkylating a mixture of toluene and $C_9{}^+$ aromatics. The latter, comprising predominantly methyl- and ethyl-substituted alkylbenzenes, was obtained by fractionating a catalytic reformate stream. The reaction provides a new route to upgrade heavy aromatics to xylenes.

While xylene disproportionation is difficult with ZSM-5, the catalyst is very effective for the disproportionation of toluene:

Table 4.14 Transalkylating a Mixture
of Toluene and $C_9{}^+$ Aromatics

Reaction Conditions: 43 atm, 370°C, 4/1 molar H_2/HC, 1 WHSV		
wt %	Feed	Product
Nonaromatics	0.2	7.4
Benzene	—	7.2
Toluene	44.4	37.6
Ethylbenzene	0.3	—
Xylenes	4.9	26.8
C_9 Aromatics	38.9	17.4
C_{10} Aromatics	9.4	1.3
$C_{11}{}^+$ Aromatics	1.9	2.3

Source: Brennan and Morrison (1976).

With larger pore zeolites the transalkylation and disproportionation of toluene and $C_9{}^+$ aromatics is easier. Chu et al. (1989) used zeolite Beta in the transalkylation of polyaromatics. Absil et al. (1991) studied the vapor phase reaction with a feedstock containing $C_9{}^+$ aromatics and benzene/toluene over a variety of steamed zeolite catalysts, including ZSM-5, MCM-22, ZSM-12, and zeolite Beta, to produce a product containing C_6–C_8 aromatics. A critical variable for a good catalyst is to have a constraint index of from 1 to about 2.5 (see Table 2.3).

Wang et al. (1990) reported the disproportionation and transalkylation of toluene and 1,2,4- and 1,3,5-trimethyl benzenes over zeolite Beta. Das et al. (1994) also studied the disproportionation of toluene and C_9 aromatics over zeolite Beta, mordenite, and ZSM-5.

Alkylation

Ethylbenzene synthesis

Zeolite catalyzed alkylation of aromatics with ethene was reported by Wise, using rare earth exchanged faujasite (Wise, 1966), and by Karge and co-workers using mordenite (Becker et al., 1973). Alkylation of benzene is accompanied by the polymerization of ethene. Hence a high molar ratio of benzene to ethene is necessary to minimize the polymerization of ethene and the formation of polyalkylated benzenes, which cause rapid catalyst deactivation (Wise, 1966).

ZSM-5 inhibits the formation of poly-alkylated benzenes normally formed in significant quantities with non–shape selective catalysts. This unique property led to more stable catalysts and improves the product selectivity far superior to all previously available catalysts. The catalyst is used in the new ethylbenzene process jointly developed by Mobil and the Badger Company.

Kresge et al. (1985) claimed that ZSM-23, one-dimensional 10-membered oxygen ring zeolite, has a better selectivity than ZSM-5 in the alkylation of benzene with ethylene. Pellet et al. (1988) reported similar reaction selectivities over some silicoalumino-phosphates such as SAPO-11, another one-dimensional 10-membered oxygen ring zeolite.

Synthesis of propylbenzenes

Alkylation of benzene with propene over HZSM-5 at below 275°C yields primarily cumene (isopropylbenzene) as expected from classical acid catalyzed alkylation.

At a 0.7 weight ratio of propene/benzene in a mixture of hexanes and heptanes, propene is completely converted, while the paraffin conversion is <5%. Most of the propene that is not consumed in alkylation goes to higher molecular weight olefins. As pressure is increased, some polyalkylated products are formed (Brennan et al., 1981) (Table 4.15).

Table 4.15 Alkylation of Benzene with Propene 3 WHSV Benzene Blend,[a] 0.5 WHSV Propene, 275°C

	Pressure, atm		
	1	28	48
Propene conv., wt %	95	99	99
Benzene alkylated, wt %	10	29	43
Products, wt %			
Methane + ethane	<0.1	<0.1	<0.1
Propane	0.4	0.3	0.3
Propene	0.7	0.1	0.1
i-Butane	0.5	0.5	0.4
n-Butane	0.2	0.3	0.5
Butenes	1.3	<0.1	<0.1
i-Pentane	0.5	0.4	0.5
n-Pentane	0.3	0.5	0.6
Pentenes	1.6	<0.1	<0.1
$C_6 + C_7$ Paraffins	69.4	68.0	65.4
C_6^+ Olefins	2.7	0.3	0.7
Benzene	17.8	14.0	11.3
Toluene	<0.1	<0.1	<0.1
C_8 Alkylbenzenes	<0.1	<0.1	<0.1
Isopropylbenzenes	3.5	3.6	10.3
n-Propylbenzenes	<0.1	<0.1	0.3
Other alkylbenzenes	1.5	7.0	9.6

[a]Benzene blend, wt %: 23% benzene, 12% *n*-hexane, 30% 2-methylpentane, 6% 2,3-dimethylbutane, 4% *n*-heptane, 20% 2,4-dimethylpentane.

Innes et al. (1992) claimed much improved selectivity and low aging rate when the alkylation of aromatics with C_2 and C_4 olefins over zeolite Beta was operated in a multi-stage liquid phase reactor. Bhandarkar and Bhatia (1994) in studying the alkylation of toluene with ethanol over HZSM-5 found that modification of the zeolite with phosphorus, boron, and magnesium yielded better selectivity (~90%) as compared to unmodified catalyst (around 50%).

Synthesis of diisopropylbenzenes

The synthesis of diisopropylbenzene (DIPB) was earlier studied by Kaeding (1989). Rapid catalyst aging was found to be the problem. Parikh et al. (1994)

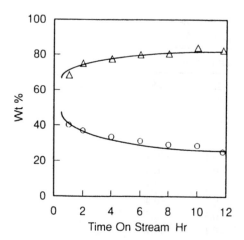

Figure 4.16 Catalytic performance of zeolite Beta: (o) cumene conversion, (Δ) DIPE selectivity. Experimental conditions: temperature = 225°C, WHSV = 5, cumene/isopropanol = 6 mol/mol. (From Parikh et al., 1994.)

found zeolite Beta to be an excellently stable catalyst. While cumene conversion decreased slowly, the selectivity to DIPB increased as shown in Figure 4.16.

Synthesis of long chain alkylbenzenes

The influence of structure on product distribution was also observed on other medium pore zeolites. Young found that ZSM-12, which has a constraint index lower than that of ZSM-5, gave unique product distributions in the alkylation of benzene (Young, 1981a) with long chain olefins. They found preferential formation of 2-phenyl alkanes as shown in Table 4.16.

$$R—CH{=}CH—CH_3 \rightleftharpoons R{=}CH_2 \rightleftharpoons R—CH{=}CH—R$$

$$CH_3—CH_2—R$$

Table 4.16 Alkylation over ZSM-12

Benzene + dodecene, 200°C, 14.6 atm[a]

Dodecene conversion, wt %	Phenyldodecane isomer distribution, %				
	2-φ	3-φ	4-φ	5-φ	6-φ
54	92	8	0	0	0

Phenol + 1-octanol, 250°C, 16.6 atm[b]

Octanol conversion, wt %	Octylphenol isomer distribution					Ratio, octylphenol to dioctylphenol
	Para		Ortho			
	2-Octyl	3-Octyl	2-Octyl	3-Octyl	Others	
100	37	41	16	0	6	135

[a]*Source*: Young (1981a).
[b]*Source*: Young (1981b).

Table 4.17 Isomer Distribution of Isopropyltoluenes

	ortho	meta	para
Reaction temperature, 230°C	5	65	30

Source: Kaeding et al. (1980).

Cymene synthesis

Alkylation of toluene with propene over ZSM-12 was studied by Kaeding and co-workers (1980). At 230–240°C and 33 atm, feeding a 6.25 molar ratio of toluene to propene at 6.1 WHSV, 90 to 95% of the propene fed was converted. The selectivity to isopropyl toluene (cymene) exceeded 95%. Table 4.17 shows the isomer distribution.

Diarylalkanes

The formation of diarylalkanes and polyalkylated benzenes during the alkylation of benzene or toluene over the medium pore zeolites is sterically hindered. However, under more severe reaction conditions, these sterically hindered reactions can take place to some extent to yield unique products only possible

Table 4.18 High Molecular Weight Products from Alkylating Toluene with Ethene

Catalyst	ZSM-5, $SiO_2/Al_2O_3 = 60$
Temperature	450°C
Pressure	1 atm
WHSV	4.5
Ethene/toluene, molar	0.2
Composition of 275–320°C product, wt %	
Diarylalkanes	85.0
C_nH_{2n-14}	
$n = 14$	15.3
$n = 15$	43.8
$n = 16$	25.9
Others	15.0
Total	100.0

Source: Sato et al. (1986).

with shape selective zeolites. For example, Sato et al. (1986) showed that in the alkylation of toluene with ethene over HZSM-5 at 450°C and 1 atm, 2.1% of the alkylated product boiled above 250°C. The fraction boiled between 275 and 320°C comprised 85% of diarylalkanes with the composition shown in Table 4.18. These compounds have been shown to be excellent electrical insulating oils.

Alkylbenzenes from paraffin/aromatic mixtures

Paraffins are normally not suitable as alkylating agents for aromatics, although in some instances low yields of alkylaromatics have been obtained with $AlCl_3$ as an acid catalyst. With medium pore zeolites, such as ZSM-5, this reaction occurs with surprising ease and good yield.

When paraffins are cracked over ZSM-5 in the presence of benzene and toluene, the aromatic molecules rapidly "scavenge" the cracked fragments (Garwood and Chen, 1980).

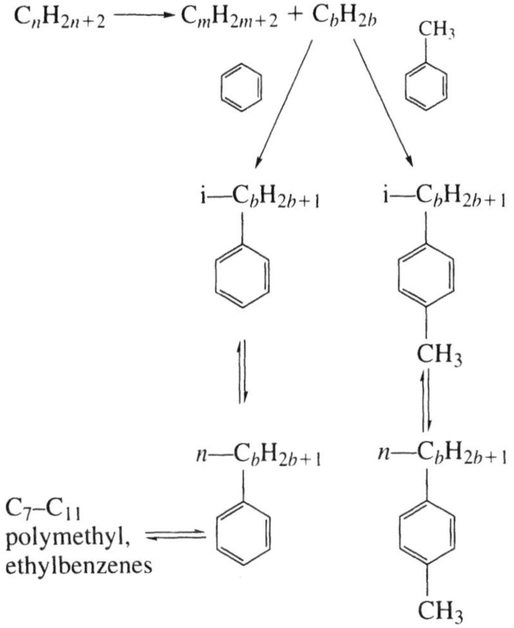

$$C_nH_{2n+2} \longrightarrow C_mH_{2m+2} + C_bH_{2b}$$

i—C_bH_{2b+1} i—C_bH_{2b+1}

CH$_3$

n—C_bH_{2b+1} n—C_bH_{2b+1}

C_7–C_{11} polymethyl, ethylbenzenes

CH$_3$

This is demonstrated by charging a mixture of *n*-octane and benzene over HZSM-5 at 49 atm and 315°C (Table 4.19) (Garwood and Chen, 1980). A carbon balance shows the fate of the octane converted (Table 4.20). Note the negligible amounts of new rings were made under these conditions. This is in contrast to the case where *n*-octane alone was converted under similar conditions (Table 4.8) and 17% of the converted octane appeared as aromatics formed by a series of reactions described earlier as M2-Forming reactions.

With benzene in the feed, the most predominant products are C_3–C_5 paraffins and propyl- and butyl-benzenes. Their formation is readily explained by an initial cracking of *n*-octane to light C_3^+ paraffins and olefins, followed by alkylation of benzene by the propene and butenes. Alkylation by pentene is sterically hindered. The primary alkylation products are those expected from classical acid-catalyzed alkylation: isopropylbenzene and *sec*-butyl benzene. Under more severe reaction conditions, they first undergo secondary isomerization reactions to form *n*-propylbenzene and *n*-butyl benzene. This is followed by dealkylation, olefin isomerization/transmutation and realkylation reactions in addition to disproportionation reactions to produce the spectrum of aromatic products.

Table 4.19   1/1 Weight Blend *n*-Octane/
Benzene over ZSM-5 at 315°C, 49 atm, 4 LHSV

	Products, wt %
Methane + ethane	<0.1
Propane	7.3
Propene	<0.1
Butanes	10.7
Pentanes	5.7
$C_6 + C_7$ Paraffins	2.4
Octane	6.4
Benzene	24.7
Toluene	0.5
C_8 Alkylbenzenes	1.3
Isopropylbenzene	8.5
n-Propylbenzene	11.2
Other C_9 alkylbenzenes	0.5
n + *sec*-Butylbenzene	7.1
tert-Butylbenzene	1.9
Other C_{10} alkylbenzenes	1.7
C_{11} + alkylbenzenes	10.1
	100.0

Source: Garwood and Chen (1980).

Table 4.20   Fate of Converted *n*-Octane

	wt %
Light paraffins	65
Alkylbenzene side chains	34
Aromatic ring carbon	<1

Source: Garwood and Chen (1980).

Cracking longer chain paraffins over ZSM-5 produces primarily C_3 and C_4 olefins which are scavenged by benzene. Thus, with *n*-hexadecane the weight percent of converted paraffin that ends up as alkyl aromatics reaches about 75% (Figure 4.17).

When both benzene and toluene are present in the reaction mixture, they are alkylated by the cracked fragments. Toluene is intrinsically more reactive for alkylation than benzene. According to the selectivity relationship developed

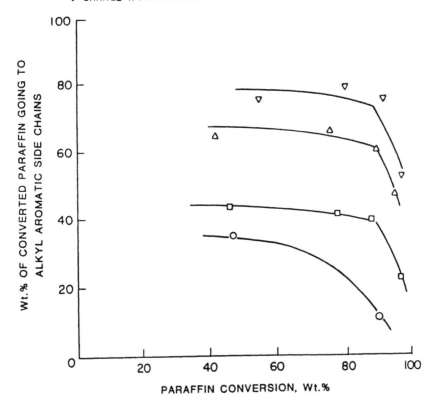

○ CHARGE 1/1 Wt. BLEND, n-PENTANE/BENZENE

□ CHARGE 1/1 Wt. BLEND, n-OCTANE/BENZENE

△ CHARGE 1/1 Wt. BLEND, n-DODECANE/BENZENE

▽ CHARGE 1/1 Wt. BLEND, n-HEXADECANE/BENZENE

Figure 4.17 Conversion of *n*-paraffins to alkyl aromatic side chains at 48 atm, 315°C; severity controlled by space velocity. (From Garwood and Chen, 1980.)

for nonstereospecific catalysts by Stock and Brown (1959) the ratio of relative rate constants, k(toluene)/k(benzene), for alkylation should be about 1.4. This relationship was found to hold for both homogeneous and heterogeneous catalyst systems (Brown and Smoot, 1956; Venuto, 1967). However, over ZSM-5, as shown by the data obtained at 35 atm (Table 4.21), the observed ratio of k(toluene)/k(benzene) is about half of that for the nonstereospecific case,

Table 4.21 Rate of Aromatics Alkylation[a]

	Temperature, °C	
	315	340
Conversion, wt %		
Benzene	25	41
Toluene	18	31
k(toluene)/k(benzene)	0.69	0.70

[a]35 atm. No hydrogen was used.
Source: Chen and Garwood (1978b).

suggesting that the channel openings in ZSM-5 either impose a sterospecific effect on the alkylation reactions or that the rate of alkylation is inhibited by the slow diffusion of some of the dialkylbenzene isomers (Chen and Garwood, 1978b).

    This interpretation is consistent with the observation that as the reaction pressure is reduced, the observed ratio of k(toluene)/k(benzene) increases toward that of the intrinsic ratio (Table 4.22) (Garwood and Chen, 1980).

Alkylation of polynuclear aromatics

Shape selectivity in the alkylation of polynuclear aromatics, such as biphenyl, naphthalene, *p*-terphenyl, and dibenzofuran, was found to form the least bulky products when using H-mordenite as the catalyst (Sugi and Toba, 1994; Sugi et al., 1994), for example, 4-isopropyl biphenyl (IPBP) and 4-4'-diisopropyl biphenyl (DIPB) for biphenyl and 2,6-diisopropyl naphthalene (2,6-DIPN) for naphthalene. ZSM-5, on the other hand, does not have enough space to ac-

Table 4.22 Effect of Pressure on the Relative Conversion (wt %) of Aromatics Alkylation[a]

	Pressure, atm	
	1	48
Benzene	24	65
Toluene	30	52
k(toluene)/k(benzene)	1.30	0.74

[a]~280°C, propene/(benzene + toluene) = 1/1 molar.
Source: Garwood and Chen (1980).

commodate 1-methyl naphthalene and dimethyl naphthalene isomers with the α-methyl group. It also can only slightly distinguish the steric difference between 2,6- and 2,7-dimethyl naphthalene.

Dealkylation of Alkylaromatics

The dealkylation of alkylaromatics over ZSM-5 as in the case of alkylation is accompanied by secondary reactions, viz., olefin isomerization/transmutation and re-alkylation reactions in addition to disproportionation and hydrogen transfer reactions to produce a spectrum of aromatics and gaseous products. Table 4.23 shows the product distribution when n-propylbenzene is dealkylated at 28 atm in 3/1 mol ratio of hydrogen to hydrocarbon.

Sato et al. (1985) showed that when the dealkylation reaction was carried out in an inert atmosphere with MgO-modified Li-ZSM-5, secondary reactions of the dealkylated olefin could be avoided. Table 4.24 compares the composition of gaseous product of H-ZSM-5 with that of the lithium-exchanged ZSM-5 when cymene was dealkylated at 1 atm and 250°C.

By adding a noble metal function to the catalyst, alkylbenzenes with C_2^+ alkyl groups can be selectively hydrodealkylated to a mixture of benzene and methyl-substituted benzenes (Bonacci and Billings, 1976; Onodera et al., 1982) with ethane as the major byproduct. For example, a feed containing 9% ethyl

Table 4.23 Dealkylation of n-Propylbenzene over HZSM-5[a]

	Temperature, °C	
	315	370
$C_1 + C_2$	0.2 (wt %)	0.4 (wt %)
C_3	4	7
C_4	6	8.5
C_5^+	4	5
Benzene	38	45
Toluene	2	4
C_8 Aromatics	4	9
C_9 Aromatics	24	9
C_{10}^+ Aromatics	18	12

[a]28 atm, 3/1 molar ratio of hydrogen to hydrocarbon.
Source: Brennan and Morrison (1976).

Table 4.24 Dealkylation of Cymene at 1 atm

Catalyst	Composition of gaseous product, mol %					
	Ethene	Propene	Propane	Butene	Pentene	Hexene
MgO modified LiZSM-5 at 500°C	0	95	0	2	1	2
HZSM-5 at 250°C	1	11	2	46	28	12

Source: Sato et al. (1985).

Table 4.25 Hydrodealkylation of Alkylbenzenes over Pt/ZSM-5 Catalyst at 1 atm and 400°C

	Feed	Product, wt %
Methane		0.01
Ethane		24.69
Ethene		0.02
Propane		0.59
Butanes		0.04
Benzene		23.62
Toluene		8.72
Ethylbenzene		3.09
Xylenes		16.55
Cumene	0.9	—
Ethyltoluene	9.0	0.45
Trimethylbenzenes	10.0	9.52
Diethylbenzenes	47.0	0.52
Ethylxylenes	33.1	12.18
Totals	100.0	100.00

Source: Onodera et al. (1982).

toluene, 10% trimethylbenzene, 47% diethylbenzene, and 33.1% ethylxylene was dealkylated at 1 atm and 400°C over a Pt/ZSM-5 catalyst to yield a mixture of BTX and ethane, as shown in Table 4.25.

Paring Reactions

The paring reaction forming light isoparaffins, particularly isobutane, from polymethylbenzenes, is described by Sullivan et al. to occur by a mechanism involving ring contraction and expansion and isomerization (Sullivan et al., 1961).

CH₃ CH₃ CH₃ CH₃ CH₃ CH₃ + H⁺ ⟶ ... ⟶ ...

To explain the observed "scrambling" of side chains around the aromatic ring to form a wide spectrum of polymethyl and ethyl benzenes from the initial alkylation products, isopropyl benzene and *sec*-butyl benzene, a "reverse" paring reaction was proposed (Chen et al., 1979a) as an alternative to the dealkylation/olefin transmutation/re-alkylation pathway. However, the relative importance of these reaction schemes remains to be investigated.

Para Selective Reactions

Molecules such as *p*-dialkylbenzenes diffuse much more rapidly into medium pore zeolites than their ortho and meta isomers do. For example, the diffusion rate of *p*-xylene in ZSM-5 is 10^5 times greater than 1,3,5-trimethylbenzene (mesitylene) (Olson et al., 1981). By adjusting the acid activity and the diffusion parameters (such as crystal size and/or chemical treatment) of the zeolite such that the reactions involving the *m*- and *o*-isomers are severely diffusion constrained, high para selectivity has been achieved in the alkylation of toluene with methanol (Chen, 1977, 1978; Chen et al., 1979b; Haag and Olson, 1978; Kaeding et al., 1981a). Similarly, selective formation of *p*-ethyl toluene was made possible by alkylating toluene with ethene (Kaeding and Young, 1978; Kaeding et al., 1982), *p*-diethylbenzene by alkylating ethylbenzene with ethene (Ishida and Nakajima, 1986), and the preferential formation of *p*-cymene (1-methyl-4-isopropylbenzene) was made possible by the alkylation of toluene with propene. A ratio of greater than 10 cymene to *n*-propyltoluene has been achieved with ZSM-5 (Dwyer and Klocke, 1977).

In addition to alkylation reactions, selective toluene disproportionation (STDP) has been achieved to yield *p*-xylene as the predominant xylene isomer (Table 4.26) (Chen et al., 1979b; Kaeding et al., 1981b). Selective dealkylation

Table 4.26   Toluene Disproportionation over Modified ZSM-5 Catalysts

	Phosphorus		Boron			Magnesium
	555°C	600°C	600°C	650°C	700°C	550°C
Pressure, atm	1	1	1	1	1	1
Toluene						
WHSV	2.1	2.1	2.8	2.8	2.8	3.5
conversion, %	0.3	0.8	5.4	9.7	13.1	10.9
Time on stream, h	0–1	1–2	1.0	1.0	1.0	1.0
Product, wt %						
benzene	0.14	0.34	2.9	5.4	8.4	4.9
toluene	99.66	99.18	94.6	90.4	86.85	89.2
p-xylene	0.10	1.19	1.85	3.1	3.9	5.2
m-xylene	0.05	0.14	0.5	0.8	0.45	0.6
o-xylene	—	0.05	0.15	0.2	0.1	0.1
C_9^+ aromatics	—	—	0	0.1	0.3	0
Total	100.00	100.00	100.00	100.0	100.00	100.0
Xylene isomers, %						
para	67	50	74	74	87	88
meta	33	37	21	20	10	10
ortho	—	13	6	6	3	2

Source: Kaeding et al. (1981b).

Table 4.27 Selective Dealkylation of *p*-Cymene over MgO
Modified LiZSM-5

Feed: 32.9% *p*-cymene, 63.6% *m*-cymene, 3.5% *o*-cymene
Pressure: 1 atm, temperature: 500°C

Contact time,	Conversion, wt %		Propene purity, mol %
	para	meta and ortho	
4.0	68.6	1.5	97
10.2	91.2	4.3	93

Source: Sato et al. (1985).

of *p*-dialkylbenzenes in a dialkylbenzene mixture has also been achieved.
(Table 4.27) (Sato et al., 1985).

Haag and co-workers (Olson and Haag, 1984) developed a correlation
between observed para selectivity in toluene disproportionation and a model
based on the interplay between diffusion and reaction. In this model, shown in
Figure 4.18, toluene diffuses into the zeolite with a diffusivity, D. It undergoes
disproportionation to benzene and a mixture of xylenes (p, m, o) which equilibrate rapidly in ZSM-5. At short residence times, the primary products ($p, m,$
o) leaving the zeolite have an isomer distribution determined by the relative rate
of isomerization, the diffusion coefficient of the respective isomers (D_p, D_m,
and D_o), and the radius of the zeolite crystal. A more rigorous mathematical
analysis based on similar premises was reported by Wei (1982), who showed
that regardless of the intrinsic product selectivity, the observed isomer distribution is altered as a result of unequal diffusion rates and isomerization rate of
the isomers. The enrichment of *p*-xylene is made possible by the much greater
diffusion coefficient of *p*-xylene. It can be shown that when the rate of isomerization is much faster than that of diffusion, viz., when the effectiveness factor
$\sqrt{kR^2/D}$ is greater than 10, the observed isomer distribution at very low conversions will contain better than 90% *p*-xylene regardless of the intrinsic isomer distribution.

At longer residence times, the primary product on re-entering the zeolite would re-isomerize to form secondary products. The composition of
the secondary product will thus be a function of conversion. Theoretical
analysis (Wei, 1982) shows that the relation between para selectivity and
conversion depends on the competitive rates of alkylation (disproportionation) and isomerization, or the ratio of $k_a/\sqrt{k_i}$. Figure 4.19 shows that by fitting

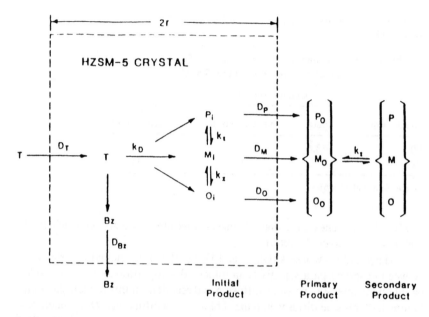

Symbols:

         D = Diffusion Coefficient
         k = Rate Constant
         r = Crystal Radius
        Bz = Benzene
         T = Toluene
     P,M,O = Para-, Meta, and Ortho-xylene

Subscripts:

         I = Isomerization
         D = Disproportionation
         i = Initial Product
         o = Primary Product

Figure 4.18   Model for STDP. (From Olson and Haag, 1984.)

the experimental data to the theoretical curves, $k_a/\sqrt{k_i}$ values of 1 and 10 are obtained for toluene disproportionation and toluene alkylation respectively.

In addition to the above examples, similar kinds of para selective reactions involving nonhydrocarbons can be found in the literature. To name a few:

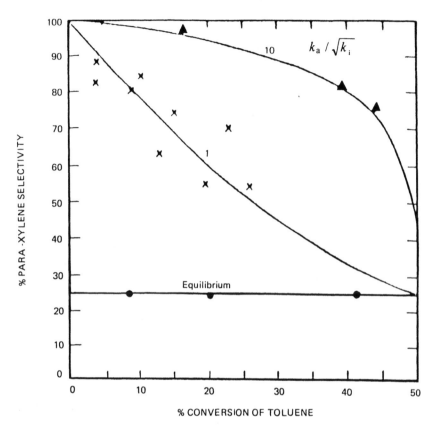

Figure 4.19 The relation between the percentage para selectivity and percentage toluene conversion in toluene disproportionation. Theoretical curves for $k_a/\sqrt{k_i}$ of 1 and 10: (●) HZSM-5 disproportionation; (X) MgZSM-5, disproportionation; (▲) PZSM-5, methanol alkylation. (From Wei, 1982.)

the synthesis of *p*-alkylanisoles (Young, 1983) and *p*-alkylphenols (Toray Industries, 1986) by alkylation of methoxybenzene and phenol, respectively, and *p*-di-halo-benzenes (Asahi Chem., 1984; Ihara Chem., 1985a) and *p*-halo-alkoxybenzenes (Ihara Chem., 1985b) by halogenation of benzene and alkoxy-benzenes, respectively.

Kwok and Jayasuriya (1994) studied the nitration of toluene with *n*-propyl nitrate using ZSM-5 as the catalyst. ZSM-5 at the high Si/Al ratio of 1000 gave the highest yield of *p*-nitrotoluene.

D. Naphthene Reactions

Naphthenes are less reactive than paraffins and olefins.

CH$_3$

△—⃝ $\longrightarrow$ C—C—C=C—C $\Longrightarrow$ C$_2$-C$_{10}$ olefins

BTX + light paraffins

      For example, the conversion of methylcyclopentane at 345°C was only 23% compared to over 80% conversion for *n*-octane. However, once the ring is opened, the resultant olefins undergo a series of M2-Forming reactions producing the wide spectrum of aromatic and paraffinic products (Table 4.28). Because of their molecular size, bicyclic naphthenes, such as decalin, are much less reactive over ZSM-5.

E. Reactions of Oxygen-Containing Compounds

Alcohols and Ethers

Hydrocarbons from methanol and dimethyl ether

The chemistry of methanol/dimethylether-to-hydrocarbons reactions over medium pore zeolites has been comprehensively reviewed by Chang (1983). There is still disagreement over the mechanism of initial C-C bond formation and whether the primary hydrocarbon from methanol/dimethyl ether is ethene or propene or both (Espinoza and Mandersloot, 1984; Chu and Chang, 1984; Wu and Kaeding, 1984). While positive identification of the primary product remains elusive, experimental data (van den Berg et al., 1980; Haag et al., 1982) lend support to the belief that ethene is the intrinsic primary hydrocarbon product. But if the rate of the secondary reaction, i.e., the autocatalytic methylation of ethene to propene reaction (Chen and Reagan, 1979a; Ono and Mori, 1981), is so much faster than the rate of desorption or diffusion of the primary product, then under diffusion/desorption disguise, propene may appear to be the primary product.

      Subsequent reactions include the redistribution of olefins via oligomerization/transmutation reactions to a thermodynamic equilibrium composition, including ethene. Radio tracer studies proved that the ethene found in the final product is not the original primary product (Dessau and LaPierre, 1982). At low

Table 4.28 Products from Methyl-
cyclopentane Cracking at 35 atm,
345°C, 1 LHSV

	wt %
Methane	<0.1
Ethane	0.3
Propane	3.6
Propene	0.1
i-Butane	1.3
n-Butane	1.3
Butenes	<0.1
i-Pentane	0.2
n-Pentane	0.3
C_6 Paraffins	5.0
C_7 Paraffins	0.5
C_8 Paraffins	0.3
Benzene	<0.1
Toluene	1.7
C_8 Alkylbenzenes	3.1
C_9 Alkylbenzenes	2.7
C_{10} Alkylbenzenes	0.7
C_{11} Alkylbenzenes	<0.1
Methylcyclopentane	77.5
Dimethylcyclopentane	0.4
Cyclohexane	0.2
Methylcyclohexane	0.3

Source: Chen et al. (1979a).

conversions, the olefin composition also correlates well with a stepwise chain-growth mechanism (Wu and Kaeding, 1984).

$$2\ CH_3OH \rightleftharpoons CH_3-O-CH_3 + H_2O$$

$$(C_2^= \rightleftharpoons C_3^= \rightleftharpoons C_4^= \rightleftharpoons C_5^= \rightleftharpoons C_6^{=+}) + H_2O$$

Cycloparaffins   Paraffins   Aromatics

Other than the effect of water and oxygenates, the chemistry of the conversion of olefins to other hydrocarbons is essentially analogous to that

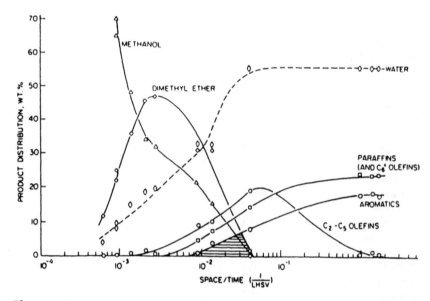

Figure 4.20 Reaction path for methanol conversion to hydrocarbons over HZSM-5 (371°C). (From Chang and Silvestri, 1977.)

described earlier. Figure 4.20 shows the reaction path for methanol conversion to hydrocarbons over HZSM-5 at 371°C (Chang and Silvestri, 1977), and the shaded area represents a region where the unconverted oxygenates and aromatics coexist leading to ring alkylation and the formation of polyalkylbenzenes. As expected, the nature of the aromatics formed from methanol is dictated by the structural effects of the zeolite.

This overlapping region may be expanded by increasing the reaction pressure and contracted by lowering the partial pressure of methanol in the feed. At higher pressures, preferential formation of durene (1,2,4,5-tetra-methylbenzene) far exceeding its equilibrium value has been observed. Thus, a new process for the manufacture of durene is possible with medium pore zeolites (Chang et al., 1974a; Fowles and Yan, 1985).

Preferential production of light olefins from methanol may be achieved in several ways, including decreasing residence time, decreasing partial pressure of the oxygenates in the feed, and operating at less than 100% conversion of the oxygenates (Chang et al., 1979).

Alternatively, reducing the acidity of the zeolite and operating at much higher temperatures (Chang et al., 1984) uncouples the olefin formation reac-

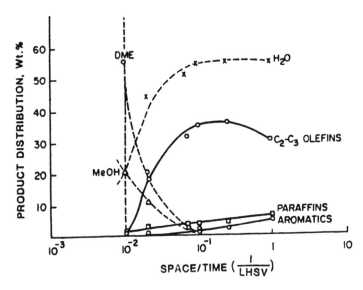

Figure 4.21   MeOH conversion over HZSM-5 (SiO$_2$/Al$_2$O$_3$ = 500) 500°C, atmospheric pressure. (From Chang et al., 1984.)

tions from the aromatization reactions. C$_2$ to C$_5$ olefin yield as high as 80% has been obtained (Figure 4.21). This approach has led to the development of Mobil's Methanol-to-Olefin (MTO) process (Gould et al., 1986).

Ethene from methanol/dimethyl ether

The selective conversion of methanol/dimethylether to ethene represents an interesting challenge to shape selective catalysis. Ethene yields over the medium pore zeolites are seldom over 10% of the total hydrocarbon product even when the reaction severity is chosen to maximize light olefins (Chang et al., 1984).

Diluting methanol with water shifts the methanol/dimethylether equilibrium toward methanol and is reported to have a dramatic effect on increasing ethene selectivity (Caesar and Morrison, 1978). Ethene selectivity as high as 28% has been observed at 10/1 molar ratio of water to methanol and 50% conversion (Figure 4.22).

A number of investigators (Chang et al., 1977; Singh and Anthony, 1979; Singh et al., 1980; Ceckiewicz, 1981; Wunder and Leupold, 1980; Inui et al.,

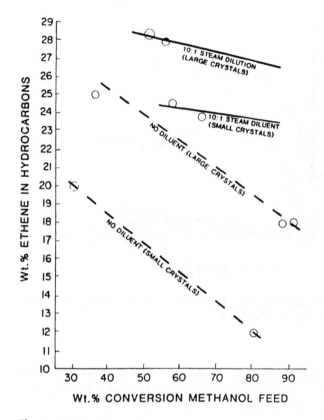

Figure 4.22 Effect of water dilution on methanol conversion to olefins. (From Caesar and Morrison, 1978.)

1981) have used small pore zeolites, such as erionite-chabazite, zeolite T, TMA-erionite/offretite, ZK-5, and ZSM-34, for methanol conversion. Representative data reported by Chang et al. are shown in Table 4.29. Although they generally produce more ethene when compared to medium pore zeolites, the reaction is always accompanied by rapid coke deactivation (Givens et al., 1978). This would be expected from structural considerations of these zeolites (Derouane, 1985).

However, Kaiser (1985) reported that with phosphate-based molecular sieves, analogs of small pore zeolites such as erionite (SAPO-17 and $ALPO_4 = 17$) and chabazite (SAPO-34 and $ALPO_4 = 17$), complete conver-

Table 4.29 Methanol Conversion over Small Pore Zeolites

	Erionite[a]	Zeolite T	Chabazite	ZK-5
Reaction conditions				
temperature, °C	370	341–378	538	538
pressure, atm	1	1	1	1
LHSV (WHSV), h^{-1}	1	(3.8)	[b]	[b]
conversion, %	9.6	11.1	100	100
Hydrocarbons, wt %				
CH_4	5.5	3.6	3.3	3.2
CH_2H_6	0.4	0.7	4.4	0
C_2H_4	36.3	45.7	25.4	21.4
C_3H_6	1.8	0	33.3	31.8
C_3H_6	39.1	30.0	21.2	13.5
C_4H_{10}	5.7	6.5 ⎫	10.4	22.6
C_4H_6	9.0	10.0 ⎭		
$C_5{}^+$	2.2	3.1	2.0	7.5

[a]De-aluminized; $SiO_2/Al_2O_3 = 16$.
[b]Pulse microreactor, 1 μL MeOH in He, 500 h^{-1} GHSV.
Source: Chang (1983).

sion of methanol was maintained for more than 6 hours. Unfortunately, operating at 100% conversion provides no information on the rate of deactivation. Further study is necessary to determine whether any difference in deactivation between these two types of molecular sieves can be attributed to either the difference in the density or the strength of their acid sites.

In any case, these studies on small pore catalysts provided a clue to selective ethene production, viz., compared to medium pore zeolites, small pore zeolites gave a higher ratio of ethene-to-$C_3{}^+$ olefins, probably the result of steric inhibition on the methanol alkylation reaction. Therefore, the pore size of the zeolite and/or diffusivity must play a role in giving high ethene selectivity.

Olson and Haag (1984) showed that it is possible to reduce the diffusivity by reducing the pore volume. In a series of experiments, they reduced zeolite sorption capacity by impregnation with magnesium oxides. Some success has been reported on increasing ethene selectivity by the addition of other inorganic oxides to the medium pore zeolites. For example, Kaeding and Butter (1980) reported improved ethene selectivity when the ZSM-5 was impregnated with phosphorus compounds. Similar effects have been observed

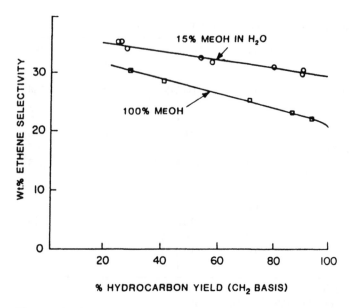

Figure 4.23 Conversion of methanol to ethene over silica-modified HZSM-5, 340–360°C. (From Rodewald, 1979.)

following impregnation of ZSM-5 with organic silicone compounds (Figure 4.23) (Rodewald, 1979). When an organic silicon-containing compound, such as dimethylsilane, was adsorbed into the pores of the zeolite, and after calcining, a reduction of zeolite sorption capacity was noted.

ZSM-5 impregnated with antimony oxide (SbO) represents another interesting approach. The zeolite can sorb more than 20 wt % of Sb_2O_3. The impregnated zeolite showed high initial ethene selectivity (Table 4.30).

However, as the antimony oxide is reduced during the reaction, it leaves the zeolite and the zeolite reverts to its unmodified form giving low ethene yield. High ethene selectivity can be maintained by cofeeding an oxidant, such as hydrogen peroxide with the methanol (Chen and Reagan, 1977).

It is interesting to note that various approaches all lead to the same upper limit of about 33% ethene at 60% conversion of methanol in a 20 wt% methanol (7:1 molar ratio of water:methanol) solution. Only by the addition of a dehydrogenation function coupled with inorganic oxide modifiers, for example, ZnPd with MgO, could the ethene selectivity be further increased to as high as 47% under the same operating conditions (Figure 4.24).

Table 4.30 Methanol Conversion over Antimony Modified ZSM-5 Catalysts

	Temperature, °C	
	300	350
Aliphatics, wt %		
H_2	0.05	0.34
CO	0.86	1.70
CO_2	0.20	0.75
Methane	0.58	1.07
Ethane	0.23	0.44
Ethene	35.90	27.05
Propane	4.55	3.32
Propene	36.92	29.79
Butanes	4.24	2.11
Butenes	13.55	10.18
C_5's	2.92	3.21
C_6's	0	1.14
$C_7{}^+$'s	0	4.24
Aromatics, wt %		
Benzene	0	0.37
Toluene	0	1.77
Xylenes	0	8.61
C_9 Alkylbenzenes	0	2.50
C_{10} Alkylbenzenes	0	1.41

Source: Butter (1976).

Hydrocarbons from ethanol and higher alcohols

The conversion of ethanol and higher alcohols to hydrocarbons is less difficult to understand than methanol because the dehydration of $C_2{}^+$ alcohols to light olefins is well known.

$$C_2H_5OH \longrightarrow C_2H_4 + H_2O$$

$$(C_2{}^= \rightleftharpoons C_3{}^= \rightleftharpoons C_4{}^= \rightleftharpoons C_5{}^= \rightleftharpoons C_6{}^{=+}) + H_2O$$

Cycloparaffins   Paraffins   Aromatics

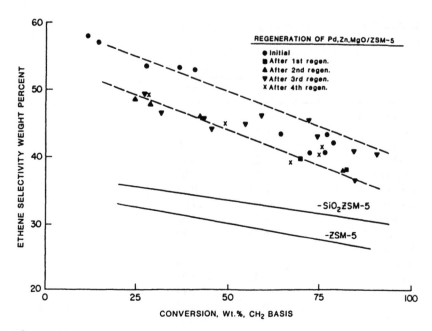

Figure 4.24 Conversion of methanol to ethene over modified ZSM-5 catalysts. (From Chen and Reagan, 1979b, 1981.)

Figure 4.25, shows the reaction pathway for ethanol conversion to hydrocarbons over HZSM-5 at 371°C (Chen, 1982; Chang, 1986). The thermochemistry of ethanol reactions is different from methanol conversion, which is highly exothermic. Although the overall heat of reaction for ethanol is less exothermic than the methanol reactions, the initial step of dehydration of higher alcohols is highly endothermic, and is followed by the highly exothermic olefinic reactions. This makes temperature control a problem in commercial reactors other than fluidized beds (Chen, 1983).

Alkylation of aromatics is again an important factor in the overall reaction scheme, as evidenced by the preponderance of ethyl-substituted side chains in the liquid product (Table 4.31).

More recently, Hutchings et al. (1994) studied the conversion of methanol and other O compounds over a larger pore zeolite Beta. Unlike other large pore zeolite, faujasite, which deactivated rapidly, zeolite Beta aged much slower and produced mainly isobutane from methanol, and light olefins only

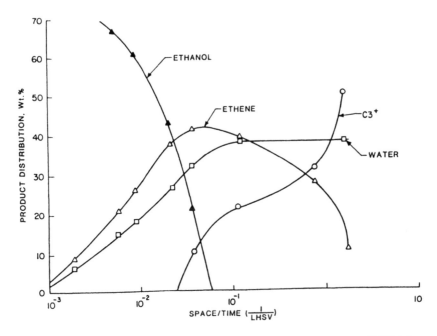

Figure 4.25 Reaction path for ethanol conversion to hydrocarbon over HZSM-5 at 371°C. (From Chen, 1982.)

from other O compounds. Bell et al. (1995) found that zeolite Beta is active at low temperatures for converting lower alkanol and C_4–C_7 tertiary olefins to high octane ether product.

Methyl tert-butyl ether (MTBE) synthesis

Methyl *tert*-butyl ether, a high octane (115-135 RON) gasoline blending component (Pecci and Floris, 1977) can be synthesized from methanol and isobutene over the medium pore zeolite. Compared to Amberlyst 15, a microreticular hydronium ion exchanged resin currently in commercial use, HZSM-5 and HZSM-11 (Chu and Kuehl, 1987) can tolerate higher temperatures, although the reaction is equilibrium limited. They also give 100% selectivity for MTBE without the need of having an excess methanol. The latter was attributed to the selective enrichment of methanol concentration in the zeolite pores.

Table 4.31 Composition of Hydrocarbon Components for Methanol and Ethanol Feedstocks

Feedstock	Methanol	Ethanol
Methane	1.14	0.28
Ethane	0.50	1.78
Ethene	0.03	0.01
Propane	5.70	12.17
Propene	0.22	0.15
n-Butane	3.24	7.75
i-Butane	8.69	10.25
Butenes	1.08	0.68
n-Pentane	1.63	3.77
i-Pentane	11.13	9.39
Pentenes	2.09	0.77
Cyclopentane	0.33	0.35
Methylcyclopentane	1.57	1.09
n-Hexane	0.83	1.15
Methylhexanes	4.61	1.77
i-Hexanes	11.72	5.62
Hexenes	1.49	0.38
n-Heptane	0.21	0.22
i-Heptanes	0.50	0.45
C_7-Olefins	1.91	0.45
Dimethyl-N5	1.82	1.14
n-Octane	0.04	0.00
$i\text{-}C_8 = P + O + N_5 + N_6$	6.73	3.01
n-Nonane	0.14	0.07
$i\text{-}C_9 = P + O + N_5 + N_6$	2.43	0.86
n-Decane	0.00	0.00
$i\text{-}C_{10} = P + O + N_5 + N_6$	0.77	0.17
Unknown (hydrocarbon aromatics)	0.11	0.30
Benzene	0.23	0.93
Toluene	2.22	6.95
Ethylbenzene	0.64	2.00
(p + m)-Xylenes	6.95	7.44
o-Xylene	1.90	2.12
Trimethylbenzenes	6.86	3.70
Methyl-ethyl-benzenes	2.74	4.37
Other C_3 alkylbenzenes	3.71	0.16
1,2,4,5-Tetramethylbenzene	3.71	0.16
1,2,3,5-Tetramethylbenzene	0.24	0.11
1,2,3,4-Tetramethylbenzene	0.11	0.29

Table 4.31 Continued

Feedstock	Methanol	Ethanol
Other C_{10} alkylbenzenes	1.65	3.27
C_{11}-Alkylbenzenes	0.54	1.53
Naphthalenes	0.07	0.35
Unknowns (all other)	1.33	2.52
(Total C_4 + gasoline)	92.41	85.61

Source: Chen (1983).

Carbonylation of methanol

Nagy (1979) in his ^{13}C-NMR investigation of the conversion of methanol over ZSM-5, in the presence of 130 Torr of CO found that CO had a negligible effect on the rate of conversion of methanol, although some carbon monoxide appeared to be incorporated into the products. The carbonylation activity of ZSM-5 was later confirmed by Feitler (1986). In the latter study, the reaction was carried out at 69 atm and 355°C, feeding CO and methanol at a molar ratio of 33:1. Methanol was converted to hydrocarbons and oxygenates with the product distribution shown in Table 4.32.

Table 4.32 Evidence of Carbonylation Activity of ZSM-5—reaction conditions: 69 atm, 355°C

HZSM-5, SiO_2/Al_2O_3	20	52	52	46
WHSV (MeOH)	2.4	0.24	0.12	2.4
MeOH conv.	6.3	23	100	33
CO/MeOH	33	33	33	—
N_2/MeOH	—	—	—	33
Selectivity,[a] *mol %*				
Acetic acid + methylacetate	23	10	5	0
Ethene	24	15	10	38
Propane	5	7	7	7
Propene	7	6	5	16
$C_4{}^+$	22	53	70	37

[a]Selectivity defined as

$$\frac{\text{Moles product} \times \text{number of MeOH-based carbons in molecules}}{\text{Total moles product carbon from MeOH}} \times 100$$

Source: Feitler (1986).

Aldehydes and Ketones

Double bond isomerization

Hölderich (1986) reported that the double bond shift of olefinic compounds containing functional groups, such as 2-ethylacrolein to *trans*-2-methyl-butenal (tiglaldehyde) can be achieved over low-acidity medium pore zeolites with excellent selectivity.

$$
\begin{array}{ccc}
\overset{\displaystyle CH_2}{\overset{\|}{R-CH_2-C-CHO}} & \rightleftharpoons & \overset{\displaystyle CH_3}{\overset{|}{R-CH=C-CHO}}
\end{array}
$$

Skeletal isomerization

Medium pore zeolites catalyze the interconversion of aldehydes and ketones by skeletal isomerization (Höldderich et al., 1985a, 1987). For example, isobutyl aldehyde is isomerized to methylethylketone at 300–500°C over ZSM-5, with better than 60% selectivity (Linstid and Koermer, 1985).

Condensation

The acid catalyzed condensation of acetone to 1,3,5-trimethylbenzene (mesitylene) is a well-known reaction. Chang and Silvestri (1977) showed that the molecular size constraint imposed by the medium pore zeolites has a decisive effect on the product formation of this reaction. Thus, instead of making the bulkier mesitylene, acetone is selectively converted to isobutene over ZSM-5 under mild reaction conditions (Chang et al., 1981).

Huang and Haag (1982) showed that by adding a metal, such as palladium to the medium pore zeolites, the reaction intermediate, mesityl oxide, can be hydrogenated to form methylisobutyl ketone (MIBK) in high yields:

$$
\underset{\text{Acetone}}{2(CH_3)_2\,CO} \xrightarrow{-H_2O} \underset{\text{Mesityl oxide}}{(CH_3)_2\,C=CHCOCH_3}
$$

$$
\xrightarrow{-H_2} \underset{\text{MIBK}}{(CH_3)_2\,CHCH_2COCH_3}
$$

Hagen (1984) reported that β-unsaturated aldehydes can be readily prepared from formaldehyde and aldehydes of the formula RCH_2CHO (where R = H, alkyl, aryl, aralkyl, cycloalkyl, or alkylaryl) via aldol condensation over medium pore zeolites.

$$R—CH_2—CHO + H_2 \; CHO \longrightarrow CH_2=CH—CHO$$

The acid catalyzed condensation of formaldehyde and isobutene, known as the Prins reaction, was studied by Chang et al. (1981) over the medium pore zeolites. Compared to non–shape selective catalysts, the molecular size constraint imposed by the medium pore zeolites was found again to inhibit the formation of bulkier cyclic condensation products such as 4,4-dimethyl-1,3-dioxane. The selective production of 2-methylbutenol provides a simpler route to isoprene:

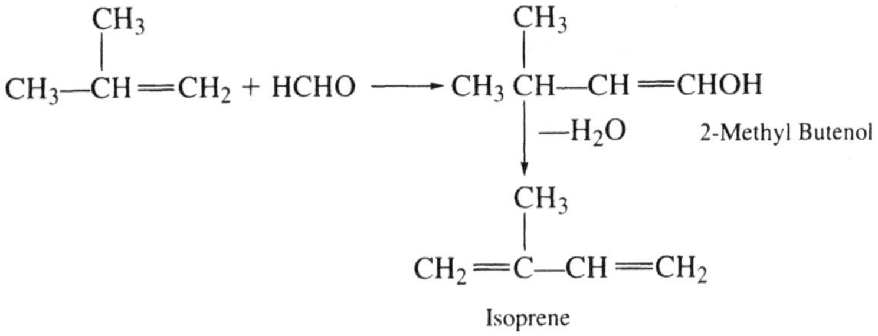

$$CH_3—\overset{\overset{\textstyle CH_3}{|}}{CH}=CH_2 + HCHO \longrightarrow CH_3 \; \overset{\overset{\textstyle CH_3}{|}}{CH}—CH=CHOH$$

$$\Big\downarrow —H_2O \qquad \text{2-Methyl Butenol}$$

$$CH_2=\overset{\overset{\textstyle CH_3}{|}}{C}—CH=CH_2$$

Isoprene

Dehydration

Hölderich et al. (1985b) reported the use of low-acidity ZSM-5 to dehydrate aldehydes such as 2-methylbutanal and isovaleraldehyde to isoprene.

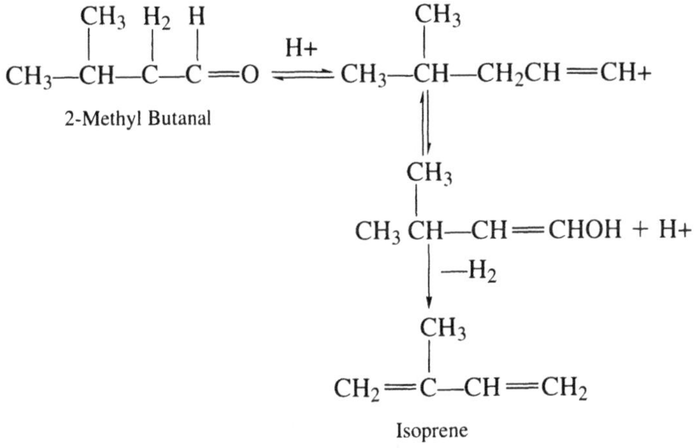

$$CH_3—\overset{\overset{\textstyle CH_3}{|}}{CH}—\overset{\overset{\textstyle H_2}{|}}{C}—\overset{\overset{\textstyle H}{|}}{C}=O \underset{}{\overset{H+}{\rightleftharpoons}} CH_3—\overset{\overset{\textstyle CH_3}{|}}{CH}—CH_2CH=CH+$$

2-Methyl Butanal

$$\Big\updownarrow$$

$$CH_3 \, \overset{\overset{\textstyle CH_3}{|}}{CH}—CH=CHOH + H+$$

$$\Big\downarrow —H_2$$

$$CH_2=\overset{\overset{\textstyle CH_3}{|}}{C}—CH=CH_2$$

Isoprene

Reactions of Acids and Esters

Conversion to hydrocarbons

The conversion of low molecular weight carboxylic acids, such as acetic acid, to hydrocarbons over HZSM-5 is accompanied by rapid catalyst deactivation and yields less hydrocarbon than that of methanol because of their low effective hydrogen content.* However, total conversion of acetic acid can be maintained indefinitely provided that the catalyst residence time is short and periodic catalyst regeneration is employed. It is interesting that decarboxylation reaction takes place extensively over ZSM-5. As a result of rejecting oxygen as CO_x, a substantial amount of hydrocarbon is formed.

Addition of methanol to acetic acid decreases the rate of catalyst deactivation. More interestingly, it shifts the oxygen rejection reaction from decarboxylation to dehydration (Figure 4.26). As a result, a synergistic effect was observed (Chang et al., 1983) in increasing the yield of total hydrocarbons and C_5^+ liquid. The C_5^+ liquid from the acid/alcohol mixtures is highly aromatic (*H/C* ratio of about 1.3) compared to a *H/C* ratio of about 2 for the liquid product from methanol.

The behavior of esters is similar to that of acid/alcohol mixtures (Table 4.33). High molecular weight fatty acids and triglycerides, on the other hand, have higher effective hydrogen content than acetic acid and are more hydrocarbonlike. Weisz et al. (1979) showed that natural products such as corn oil, peanut oil, castor oil, and jojoba oil can be converted over ZSM-5 to a mixture of paraffins, olefins, and aromatics, with a product distribution similar to that from methanol (Figure 4.27).

Fries rearrangement

The intramolecular transformation of phenyl acetate into hydroxyacetophenones, known as the Fries rearrangement, was studied by Pouilloux et al. (1987) over acid catalysts at 400°C. Compared to large pore zeolites, such as HY, the HZSM-5 catalyst was more stable and remained active for hours. The para-to-

---

*For compounds containing oxygen, their effective hydrogen content, or H/C ratio, can be expressed as follows:

$$\left(\frac{H}{C}\right)_{\text{eff}} = \frac{H - 2(O)(w)}{C - O(cm) - 0.5(O)(cd)}$$

where H, C, and O are the number of hydrogen, carbon, and oxygen atoms in the empirical formula of the feed, respectively, and w, cm, and cd are the fractions of the oxygen rejected as water, carbon monoxide, and carbon dioxide, respectively. Assuming all the oxygen to be rejected as water, then

$$\left(\frac{H}{C}\right)_{\text{eff}} = \frac{H - 2(O)}{C}$$

For example, acetic acid has an effective *H/C* ratio of zero (Haag et al., 1980).

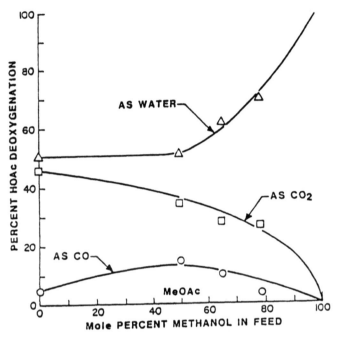

Figure 4.26 Effect of methanol on acetic acid deoxygenation. (From Chang et al., 1983.)

ortho ratio of hydroxyacetophenone produced over HZSM-5 was about 10 times higher than that over non–shape selective catalysts.

Phenyl acetate   Hydroxyacetophenone

Carbohydrates

When carbohydrates are thermally decomposed, water molecules are driven off leaving a carbon residue, thus removal of oxygen in the form of water

Table 4.33 Conversion of Ester and Acid/Alcohol Mixture over HZSM-5 (410°C, 1 atm, 1.0–1.1 WHSV, 20 min reaction intervals)

	Methyl acetate	1.9/1 (molar) Methanol/ acetic acid
Total conversion	89.4	>91
Conversion to non-oxygenates	86.1	90.4
Products (wt % of charge)		
CO	6.2	2.1
CO_2	17.6	9.4
H_2O	21.5	45.3
Oxygenates	13.9	9.6
C_4^- Hydrocarbon gas	6.0	7.9
C_5^+ Liquid hydrocarbon	32.1	24.9
Total hydrocarbons	38.1	32.8
Coke	2.7	0.8
Wt % of hydrocarbon		
Methane + ethane	5.6	7.2
Propane	0.7	0.4
Propene	6.7	13.8
Isobutane	0.3	0.4
n-Butane	0.0	0.1
Butenes	1.4	1.6
Subtotal	14.7	23.5
C_5^+ (gasoline)	78.7	74.1
Coke	6.6	2.4
H/C of C_5^+	1.3	~1.4

Source: Chang et al. (1983).

molecules does not produce hydrocarbons. Oxygen must be removed as carbon monoxide or dioxide in order to enrich product hydrogen.

With short catalyst residence time and frequent regeneration, hydrocarbons are produced directly from carbohydrates over HZSM-5. However, the hydrocarbon yield appeared to decrease with increasing molecular size of the sugar molecule, suggesting that the reactions leading to hydrocarbon formation depend on getting the sugar molecules into the intracrystalline pores of the zeolite (Table 4.34). Figure 4.28 reveals that between 9 and 21.8% of the carbon in the carbohydrate fed to the reactor is converted to hydrocarbons.

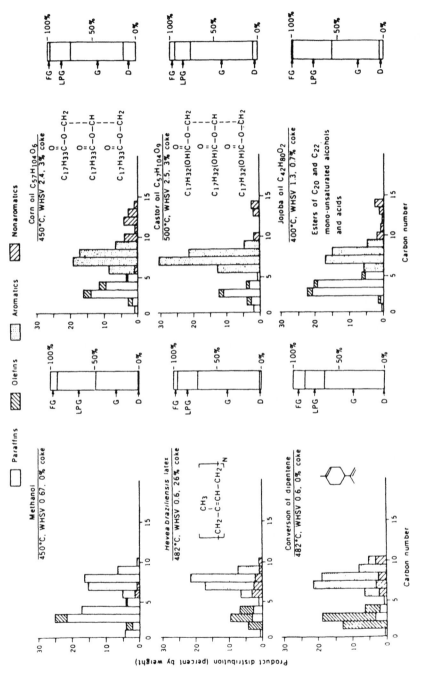

Figure 4.27 Product spectra from catalytic conversion of methanol, and of various biomass constituents. The abbreviations FG, LPG, G, and D are for fuel gas, liquid petroleum gas, gasoline, and light distillate, respectively; WHSV is for weight-hourly space velocity. (From Weisz et al., 1979.)

125

Table 4.34 Direct Conversion of Carbohydrates at 510°C

	50% xylose solution	50% glucose solution	70% starch solution	50% sucrose solution
WHSV	1.9	2.1	2.1	2.1
Product, wt %				
Hydrocarbons	10.0	8.1	8.9	4.4
CO	33.3	18.9	16.8	32.8
CO_2	3.7	3.7	1.5	8.6
Coke	16.8	24.9	30.4	23.8
H_2O	36.2	44.4	42.4	33.4

Source: Chen et al. (1986).

Phenols

Synthesis of phenol

Phenol is coproduced by the cleavage of cumene hydroperoxide over a H-Beta catalyst at 100°C (Chang and Pelrine, 1984). Typical data are shown in Table 4.35.

Isomerization

Similar to the isomerization of alkylbenzenes, the isomerization of substituted phenols having similar molecular sizes as that of xylenes, such as

Table 4.35 Cumene Hydroperoxide (CHP) Conversion over Zeolite Beta (reaction conditions: 107°C, 38 LHSV)

	Feed	Effluent
Cumene	9.0	11.5
2-Phenyl-2-propanol	6.0	1.5
CHP	82.6	22.3
α-Methylstyrene	0.3	3.2
Acetophenone	2.1	2.1
Acetone		22.5
Phenol		36.9
Totals	100.0	100.0

Source: Chang and Pelrine (1984).

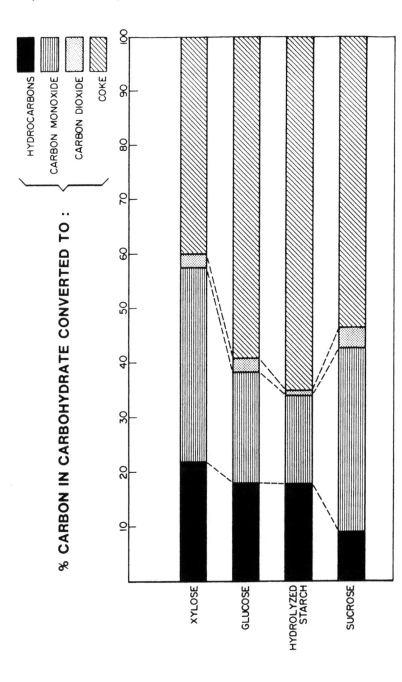

Figure 4.28 Product distribution of carbohydrates at 510°C, 2 WHSV (weight hourly space velocity), 1 atm. (From Chen et al., 1986.)

Table 4.36   Isomerization of *o*-Cresol

	ZSM-5	ZSM-12
Temperature, °C	380	420
LHSV	2.8	2.5
Pressure, atm	60	60
Product, wt %		
Phenol	1.5	3.9
o-Cresol	59.0	53.4
m-Cresol	28.3	27.7
p-Cresol	10.1	11.1
Xylenol	1.1	3.7
Trimethylphenols	0.0	0.2

Source: Keim et al. (1981).

mono-alkylphenols, has also been reported. Keim et al. (1981) showed that the isomerization of cresols proceed over ZSM-5, ZSM-11, and ZSM-12.

The isomerization reaction can be carried out readily at 380–420°C. Selectivity in excess of 95% is achieved. Compared to non–shape selective catalysts, medium pore zeolites inhibit the undesirable disproportionation reactions which lead to the formation of phenol and higher alkylated products. Typical experimental data on the isomerization of *o*-cresol are shown in Table 4.36. The equation for isomerization of cresols follows:

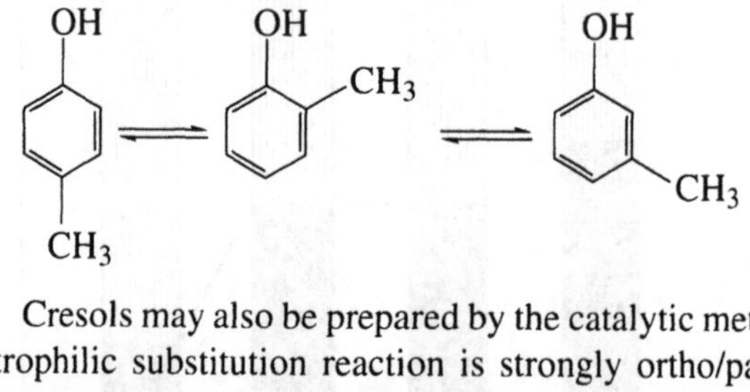

Cresols may also be prepared by the catalytic methylation of phenol. The electrophilic substitution reaction is strongly ortho/para directing. The meta isomer can be made by subsequent isomerization.

Synthesis of long chain alkylphenols

As discussed earlier for the alkylation of benzene with long chain olefins over ZSM-12, Young (1981b) also found that this zeolite catalyzes the preferential formation of *para*-mono-alkylphenols when phenol is alkylated with long chain olefins; their data are also shown in Table 4.15.

$$R-CH=CH-CH_3 \rightleftharpoons R=CH_2 \rightleftharpoons R-CH=CH-R$$

Butylation of phenol over fairly large pore zeolites, such as zeolite Beta and ZSM-12, with butene or isobutanol (Chang and Hellring, 1994) was found to be highly para selective. The primary product is *para-tert*-butyl phenol.

Acylation of phenol

The acylation of phenol by acetic acid over zeolite ZSM-5 is unexpectedly oriented toward *o*-hydroxyacetophenone by the dealumination of the external surface sites of the crystals (Neves et al., 1994). They postulated that the *o*-hydroxyacetophenone was formed through C-acylation of phenol with acetic acid due to a pronounced stabilization of the transition state. Phenyl acetate and *o*-hydroxyacetophenone are the primary products, and the formation of *p*-hydroxyacetophenone, which does not occur through phenol acylation, is formed by the autoacylation of phenyl acetate. The observed para selectivity is due to steric hindrance to the approach of the acetyl group in the ortho position of phenyl acetate.

F. Reactions of Nitrogen-Containing Compounds

Alkylamines

Shen et al. (1994) studied the amination of ethanol over a variety of zeolite catalysts. They found zeolite Beta to be an excellent catalyst to selectively produce ethylamine. Table 4.37 shows the result at 410°C, 8 WHSV, and ammonia/ethanol = 3/1 mol/mol.

Dialkylanilines

Reaction between aniline and an excess of lower alcohols or ethers at 300°C over ZSM-5 yielded *N, N*-dialkylanilines (Buysch et al. 1992). Conversion and dialkyl product selectivity increased with reaction pressure.

Table 4.37 The effects of catalysts on the amination of ethanol (experimental conditions: 410°C, 8 WHSV and ammonia/ethanol = 3/1 mol/mol)

Catalyst	Ethanol conversion, %	Ethylamine, %	Ethene, %
H-Beta	30.0	80.3	19.6
H-Mordenite	51.5	53.3	46.7
H-ZSM-5	57.5	42.8	57.2
H-Y	96.1	34.0	67.0
γ-Alumina	98.0	28.7	71.3

Source: Shen et al. (1994).

The alkylation of aniline with dimethyl carbonate to *N*-methylaniline over a faujasite catalyst was studied by Fu and Ono (1993). Hari Prasad Rao et al. (1995) reported 100% selectivity to *N*-alkylation (NMA + NNDMA) with low acidity Cs-Beta and Na-EMT catalysts.

Aromatic Amines

Reactions between phenols, alicyclic alcohols or ketones, such as cyclohexanol or cyclohexanone, and ammonia or alkyl amines over ZSM-5 yields selected aromatic amines such as aniline and *p*-toluidine (Chang and Lang, 1983, 1984). Such bulky products as diphenylamine and carbazole are apparently excluded.

Aniline can also be synthesized by nucleophilic substitution of chlorobenzene with ammonia. Copper-exchanged zeolites were found by Burgers and van Bekkum (1994) to be active for this reaction, and acidity of the catalysts enhances the substitution rate. Cu-ZSM-5 and Cu-mordenite

Toluidine Isomerization on HZSM5

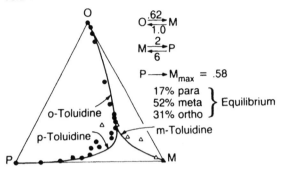

Figure 4.29 Toluidine isomerization on HZSM-5. (From Weigert, 1986.)

were found to be highly selective excluding the formation of diphenylamine and benzene.

Aromatic amines can also undergo isomerization reactions over the medium pore zeolites. Isomerization of toluidines proceeds readily over HZSM-5 with little or no aging (Eichler et al., 1985). Weigert (1986) showed that toluidines are isomerized over ZSM-5 at 300 to 400°C and atmospheric pressure to an equilibrium mixture (Figure 4.29).

Eichler et al. (1986) showed that 2-ethyl aniline is isomerized to 3-ethyl aniline with no olefinic byproducts. Interestingly, consistent with the critical dimension of the isomers, the isomerization of dimethylanilines (Weigert, 1986, 1987) occurs only with the interconversion among three isomers (2,4-, 2,5-, and 3,4-) while the other three isomers (2,3-, 2,6-, and 3,5-) are neither formed nor reactive over ZSM-5.

Pyridines

Mixtures of pyridine and alkyl pyridines can be synthesized with excellent yields by reacting carbonyl compound such as acetaldehyde, formaldehyde, and acetone with ammonia (Chang and Lang, 1980) over HZSM-5. The reaction takes place at 425°C and atmospheric pressure. The selectivity for pyridine production can be enhanced when the reaction is carried out in a fluid bed with short residence times (Feitler et al., 1987). Reacting ethanol with ammonia to form mixed pyridine bases over medium pore zeolites was reported by van der Gaag et al. (1986):

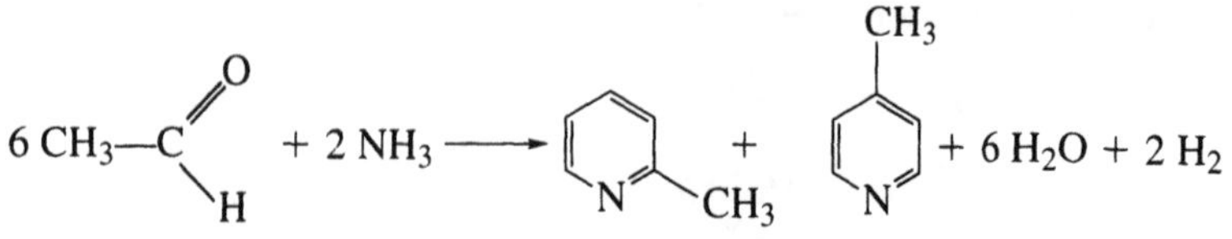

$$6 \; CH_3-\overset{O}{\underset{H}{C}} \; + \; 2 \; NH_3 \longrightarrow \; \text{α-picoline} \; + \; \text{γ-picoline} \; + \; 6 \; H_2O \; + \; 2 \; H_2$$

α-picoline   γ-picoline

$$4 \; CH_3-\overset{O}{\underset{H}{C}} \; + \; 3 \; H-\overset{O}{\underset{H}{C}} \; + \; 2 \; NH_3$$

$$\longrightarrow \; \text{pyridine} \; + \; \text{β-picoline} \; + \; 7 \; H_2O \; + \; H_2$$

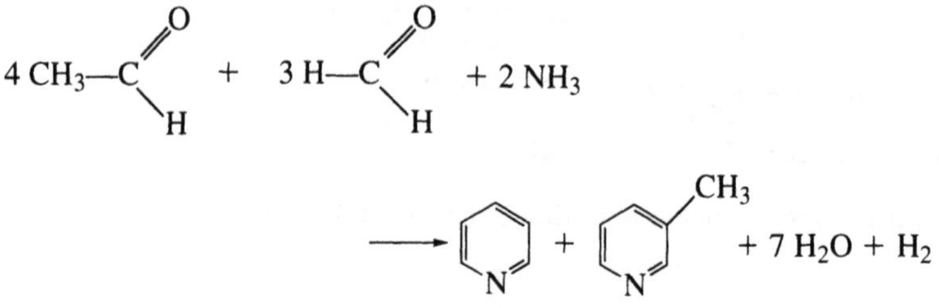

pyridine   β-picoline

Compared to larger pore zeolites, such as mordenite, the HZSM-5 catalyst was more stable and produced much less polyalkyl pyridines and heavy products, as expected (Table 4.38).

α-Picoline can also be selectively produced from aniline or phenol with ammonia over the ZSM-5 catalyst (Chang and Perkins, 1983). By increasing the ratio of ammonia to phenols fed, the reaction can be directed toward the selective production of α-picoline, without making either the β or the γ isomers.

Hölderich et al. (1992) claimed that substituted pyridines of the formula:

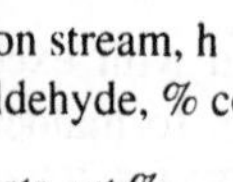

Table 4.38 Reaction of Acetaldehyde and Ammonia

Catalyst	HZSM-5		H-mordenite	
Time on stream, h	0.5	3.0	0.5	3.0
Acetaldehyde, % conversion	80	60	65	4
Products, wt %				
Light products	4.0	4.5	9.5	9.3
Pyridine	44.9	45.3	28.5	9.0
Picolines	28.1	24.5	25.2	13.9
Lutidines	17.0	19.6	27.3	33.8
Heavy products	5.9	6.1	9.5	34.0

Source: Chang and Lang (1980).

where R is an alkyl group of 1 to 20 carbon atoms, cycloalkyl, aryl, or aralkyl, are prepared by reacting acrolein and alkanals with ammonia over borosilicate or aluminosilicate of the pentasil group catalyst at 350–400°C to yield ethyl pyridine, butyl pyridine, hexyl pyridine, phenyl pyridine, and benzyl pyridine.

Le Blanc et al. (1986) reported that 2-amino-6-alkyl pyridine can be selectively produced by isomerizing 1,3-diaminobenzene or reacting, in the presence of ammonia, 3-amino phenol or 1,3-dihydroxybenzene over a medium pore zeolite. The reaction, specific to the meta isomer, is carried out at about 400°C under pressure.

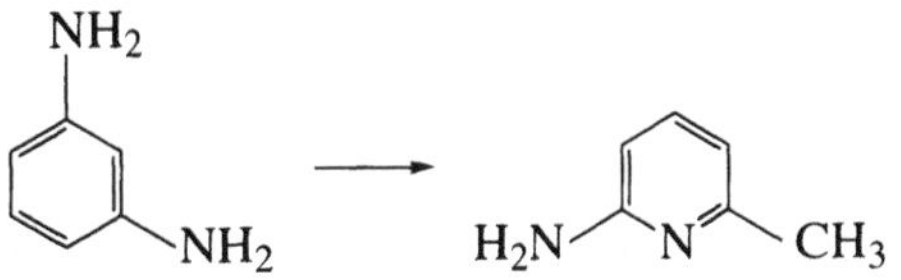

diaminophenol   2-amino-6-methyl
pyridine

Nitriles

Chang et al. (1980) also found that by reacting a mixture of formaldehyde, alcohol, and ammonia or alkyl amines over ZSM-5, alkyl nitriles and alkyl-aminonitriles are produced. Ammoxidation of toluene to benzonitrile over a copper-exchanged HZSM-5 was reported by Oudejans et al. (1984). These reactions are given:

Alkyl nitriles, alkyl amino-nitriles

$$HCHO + CH_3OH + NH_3 \longrightarrow CH_3CN + \quad + CH_3{-}NH{-}CH_2CN$$

N-methyl amino acetonitrile

$$+ (CH_3)_2{-}N{-}CH_2CN + H_2O$$

N,N-dimethyl amino acetonitrile

Aceto-nitriles

$$3\ HCHO + NH_3 \longrightarrow CH_3CN + H_2O$$

Benzonitrile

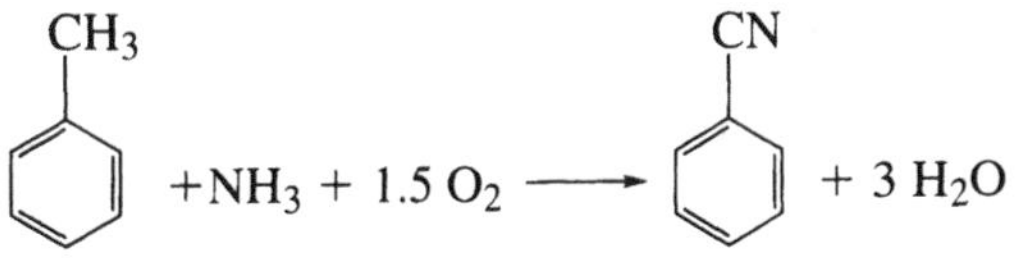

Weigert (1986) showed that ZSM-5 catalyzes methyl migration in toluonitriles and dimethylbenzonitriles by intramolecular 1,2-shifts. Only the three dimethylbenzonitriles with the 1,2,4-substitute pattern are small enough to take place in the shape selective reactions.

Caprolactam

The Beckmann rearrangement of cyclohexanone oxime to ε-caprolactam when carried out over an acid catalyst such as zeolite Y (Landis and Venuto, 1966; Aucejo et al., 1986) is accompanied by rapid coking and less selectivity. Bell and Chang (1982) showed that with medium pore zeolites, the catalysts remain active for hours. Compared to using H_2SO_4, the solid acid catalyst eliminates the byproduct, ammonium sulfate.

Bell and Haag (1990) found that by steaming ZSM-5 or ZSM-11 at 538°C for 7 hours to reduce their acidity to less than 1 alpha (see Section III of Chapter 2), selectivity and aging properties of the catalyst were significantly improved. Further improvement by eliminating surface activity with oxalic acid was demonstrated by Apelian et al. (1994).

Yashima and Komatsu (1994) claimed the use of a zeolite having closed pores as a solid catalyst in the gas phase for this reaction.

cyclohexanone oxime   caprolactam

Cyanamide

The synthesis of cyanamide from carbon dioxide or urea and ammonia is usually accompanied by the formation of undesirable byproducts of oligomerization reactions, such as melamine, a cyanamide trimer. van Hardeveld et al. (1986) reported that these secondary reactions can be avoided by using a zeolite catalyst with pores openings less than 8 Å. The reaction is carried out at about 400°C with isocyanic acid or urea, in the presence of ammonia.

$$2 \, HN{=}C{=}O \longrightarrow H_2N{-}CN + CO_2$$

isocyanic   cyanamide
acid

G. Reactions of Other Nonhydrocarbons

Conversion to Hydrocarbons

Similar to methanol and many other oxygen-containing compounds, many halogen-, sulfur-, and nitrogen-containing compounds can be converted to hydrocarbons readily over the medium pore zeolites (Chang et al., 1975b). Butter et al. (1975) reported the conversion of aliphatic mercaptans, aliphatic sulfides, aliphatic halides, and aliphatic amines to a mixture of gaseous and liquid hydrocarbons. Instead of producing water as the major byproduct, hydrogen is rejected alone with the halogen, sulfur, or nitrogen in the molecule.

As in the case of oxygen-containing compounds, the hydrocarbon yields and the rate of catalyst deactivation depend on the effective hydrogen content of the feed defined as

$$\left(\frac{H}{C}\right)_{\text{eff}} = \frac{H - 2(O + S + X) - 3N}{C}$$

where H, C, S, X, and N are the number of hydrogen, carbon, oxygen, sulfur, halogen, and nitrogen atoms in the empirical formula of the feed, respectively, assuming all the oxygen to be rejected as water.

Chang and Perkins (1987) also showed that with low effective hydrogen feeds, such as chloroform, $CHCl_3$ can react with methane over ZSM-5 to form methyl chloride, which in turn can be converted to higher molecular weight hydrocarbons over the same catalyst.

Isomerization

The isomerization activity of medium pore zeolites has also been studied over a variety of nonhydrocarbon aromatics and heterocyclic compounds, including dichlorotoluenes (Litterer, 1985; Toray Industries, 1985), halogenated thiophenes (Eichler and Leupold, 1986) and halogenated phenols (Baltes and Leupold, 1986). The latter reported that the isomerization of 2-chlorophenol and/or 4-chlorophenol to an equilibrium mixture of monochlorophenols takes place readily over HZSM-5.

Isomerization of *o*-dichlorobenzene over a variety of zeolites was studied by Coq et al. (1990). All the catalysts tested were deactivated by the formation of a coke deposit. The order of stability of the zeolites was as follows: H-ZSM-5 > H-Beta, H-offretite > H-mordenite. The activity of the zeolites ranked according to H-ZSM-5 > H-mordenite > H-Beta > H-offretite. H-ZSM-5 gave the highest yield of *p*-dichlorobenzene.

Isomerization of chlorophenols

Halogenation of Aromatics

The chlorination of benzene over zeolite K-L catalyst was reported by Naka-mura et al. (1992). With dichloromethane the liquid phase reaction gave very high para selectivity for 1,4-dichlorobenzene. Singh et al. (1994) screened a number of zeolites for the liquid phase chlorination of toluene. They found the ratio of *p*-chlorotoluene to *o*-chlorotoluene to vary with the zeolites as follows:

$$\text{K-L} > \text{K-Beta} \sim \text{K-Y} \cong \text{K-mordenite} \cong \text{K-ZSM-5} > \text{None}$$

They concluded that their data cannot be explained by the geometry-related shape selectivity alone.

H. Acid Catalyzed Hydrogenation Activity

The controversial question of whether or not acid sites in zeolites catalyze hydrogenation reactions (Minachev et al., 1968; Chen and Garwood, 1977) has received some renewed interest. Dadalla et al. (1983) reconfirmed the earlier studies on mordenite (Minachev et al., 1971, 1972) showing that Zeolon-500, a synthetic mordenite, is active for the hydrogenation of ethene. Chang and Socha (1984), in their study on the conversion of synthesis gas to hydrocarbons, found that HZSM-5 has small but significant activity for hydrogenating carbon monoxide to light saturated hydrocarbons. The role of molecular hydrogen in acid catalyzed reactions was confirmed by Haag and Dessau (1984). They showed that the restricted pore geometry of the medium pore zeolites promotes a monomolecular paraffin cracking reaction which produces molecular hydrogen. Sano et al. (1987) also demonstrated the hydrogenation of benzene over HZSM-5.

III. SELECTIVE OXIDATION WITH TITANIUM ZEOLITES

By the isomorphous substitution of silicon by titanium in ZSM-5, a titanium-containing zeolite, TS-1, a novel derivative of ZSM-5, was first synthesized by the researchers at ENI of Italy (Taramasso et al., 1983; Perego et al., 1986).

They demonstrated its remarkable activity and selectivity for the H_2O_2 oxidation of alkenes and aromatics as summarized by Notari (1987):

$$R-C = C + H_2O_2 \longrightarrow R-C-C + H_2O$$

$$R-CH_2-OH + H_2O_2 \longrightarrow R-CHO + H_2O$$

TS-1 has found industrial application in the production of hydroquinone and catechol.

TS-2, an analog of ZSM-11, was synthesized by Clerici (1991) and reported by Reddy et al. (1991) to be active for selective H_2O_2 epoxidation of alkanes and alkenes. Shape selectivity as a function of pore size in the epoxidation of cyclohexene and hexene was demonstrated (Tatsumi et al., 1991) with a series of titanium-containing zeolites, including TS-1, Na-mordenite, and dealuminized USY.

The synthesis of Ti-Beta (Camblor et al. 1993) further extended the catalytic chemistry to larger cycloalkane and cycloalkene molecules using hydrogen peroxide and *tert*-butyl hydroperoxide (Corma et al., 1994, 1995). Sato et al. (1994) studied the epoxidation of 1-octene over Ti-Beta with *tert*-butyl hydroperoxide and found the selectivity to epoxide increased to 92–100% when the Brønsted acid sites were alkali or alkaline earth ion exchanged.

REFERENCES

Abdul Hamid, S. B., E. G. Derouane, G. Demortier, J. Riga, and M. A. Yarmo, Appl. Catal. A, *108*, 83 (1994).

Absil, R. P., S. Han, D. O. Marler, S. S. Shihabi, J. C. Vartuli, and P. Varghese, U.S. Pat. 5,030,787, Jul. 9, 1991.

Alberty, R. A., J. Phys. Chem. *87*, 4999 (1983).

Apelian, M. R., W. K. Bell, A. S. Fung, W. O. Haag, and C. Venkat, U.S. Pat. 5,292,880, Mar. 8, 1994.

Asahi Chem., Japan Pat. 59-219241, Dec. 9, 1984.

Aucejo, A., M. C. Burguet, A. Corma, and V. Fornes, Appl. Catal. *22*, 187 (1986).

Baltes, H. and E. I. Leupold, U.S. Pat. 4,568,777, Feb. 4, 1986.

Becker, K. A., H. G. Karge, and W. D. Streubel, J. Catal. *28*, 403 (1973).

Bell, W. K. and C. D. Chang, U.S. Pat. 4,359,421, Nov. 16, 1982.

Bell, W. K. and W. O. Haag, U.S. Pat. 4,927,924, May 22, 1990.

Bell, W. K., W. O. Haag, and D. O. Marler, U.S. Pat. 4,962,239, Oct. 9, 1990.

Bell, W. K., S. H. Brown, M. N. Harandi, and J. C. Trewella, U.S. Pat. 5,414,146, May 9, 1995.

Bhandarkar, V. and S. Bhatia, Zeolites *14*, 439 (1994).

Bonacci, J. C. and R. P. Billings, U.S. Pat. 3,948,758, Apr. 6, 1976.

Brennan, J. A. and R. A. Morrison, U.S. Pat. 3,945,913, Mar. 23, 1976.

Brennan, J. A., W. E. Garwood, S. Yurchak, and W. Lee, Proc. Int. Seminar on Alternate Fuels, Liege, Belgium, May 1981, A. Germain, ed., p. 19.1 (1981).

Brown, H. C. and C. R. Smoot, J. Am. Chem. Soc. *78*, 6255 (1956).

Burgers, M. H. W. and H. Van Bekkum, J. Catal. *148*, 68 (1994).

Butter, S. A., U.S. Pat. 3,979,472, Sep. 7, 1976.

Butter, S. A., A. T. Jurewicz, and W. W. Kaeding, U.S. Pat. 3,894,107, Jul. 8, 1975.

Buyan, F. M., N. Y. Chen, R. B. LaPierre, D. A. Pappal, R. D. Partridge, and S. F. Wong, Hydrocarbon Tech. Intern. Autumn Quarterly 1995, p. 14.

Buysch, H.-J., H. Pelster, L. Puppe, and P. Wimmer, U.S. Pat. 5,087,754, Feb. 11, 1992.

Caesar, P. D. and R. A. Morrison, U.S. Pat. 4,083,889, Apr. 11, 1978.

Camblor, M. A., A. Corma, and J. Perez-Pariente, ES Pat. 2,037,596, Jun. 16, 1993.

Ceckiewicz, S., J. Chem. Soc. Faraday Trans. *177*, 269 (1981).

Chang, C. D., Catal. Rev.-Sci. Eng. *25*, 1 (1983); *Hydrocarbons from Methanol*, Marcel Dekker, New York, 1983.

Chang, C. D. and S. D. Hellring, U.S. Pat. 4,620,044, Oct. 28, 1986.

Chang, C. D. and S. D. Hellring, U.S. Pat, 5,288,927, Feb. 22, 1994.

Chang, C. D. and W. H. Lang, U.S. Pat. 4,220,783, Sep. 2, 1980.

Chang, C. D. and W. H. Lang, U.S. Pat. 4,380,669, Apr. 19, 1983.

Chang, C. D. and W. H. Lang, U.S. Pat. 4,434,299, Feb. 28, 1984.

Chang, C. D. and N. J. Morgan, U.S. Pat. 4,214,107, Jul. 22, 1980.

Chang, C. D. and B. P. Pelrine, U.S. Pat. 4,490,565, Dec. 25, 1984.

Chang, C. D. and P. D. Perkins, U.S. Pats. 4,388,461, Jun. 14, 1983; 4,395,554, Jul. 26, 1983.

Chang, C. D. and P. D. Perkins, U.S. Pat. 4,654,449, Mar. 31, 1987.

Chang, C. D. and A. J. Silvestri, J. Catal. *47*, 249 (1977).

Chang, C. D. and R. F. Socha, U.S. Pat. 4,472,535, Sep. 18, 1984.

Chang, C. D., A. J. Silvestri, and R. L. Smith, U.S. Pat. 3,894,105, Jul. 8, 1975a.

Chang, C. D., W. H. Lang, and A. J. Silvestri, U.S. Pat. 3,894,104, Jul. 8, 1975b.

Chang, C. D., W. H. Lang, and A. J. Silvestri, U.S. Pat, 4,062,905, Dec. 13, 1977.

Chang, C. D., W. H. Lang, and R. L. Smith, J. Catal. *56*, 169 (1979).

Chang, C. D., W. H. Lang, and R. B. LaPierre, U.S. Pat. 4,231,955, Nov. 4, 1980.

Chang, C. D., W. H. Lang, and W. K. Bell, *Catalysis of Organic Reactions*, W. R. Moser, ed., Marcel Dekker, New York, p. 73, 1981.

Chang, C. D., N. Y. Chen. L. R. Koenig, and D. E. Walsh, Am. Chem. Soc. Div. Fuel Chem. Preprints *28*(2), 146 (1983).

Chang, C. D., C. T.-W. Chu, and R. F. Socha, J. Catal. *86*, 289 (1984).

Chang, C. D., C. T. Chu, T. F. Degnan, and S. B. McCullen, U.S. Pat. 5,061,466, Oct. 29, 1991.

Chang, C. L., "Dehydration of Ethanol over ZSM-5." doctor dissertation, Michigan State University, 1985; Diss. Abtr. Int. B, *47*(2), 711 (1986).

Chen, N. Y., U.S. Pat. 4,002,697, Jan. 11, 1977.

Chen, N. Y., U.S. Pat. 4,100,215, Jul. 11, 1978.

Chen, N. Y., "Ethanol Fuel from Biomass," paper presented at the Tri-State Catalysis Club, Lexington, Ky., Oct. 20, 1982.

Chen, N. Y., Chemtech *13*, 488 (1983).

Chen, N. Y. and W. E. Garwood, J. Catal. *50*, 252 (1977).

Chen, N. Y. and W. E. Garwood, J. Catal. *53*, 284 (1978a).

Chen, N. Y. and W. E. Garwood, J. Catal. *52*, 453 (1978b).

Chen, N. Y. and W. E. Garwood, Ind. Eng. Chem., Prod. Res. Devel. *25*, 641 (1986).

Chen, N. Y. and W. J. Reagan, U.S. Pat., 4,049,735, Sep. 20, 1977.

Chen, N. Y. and W. J. Reagan, J. Catal. *59*, 123 (1979a).

Chen, N. Y. and W. J. Reagan, U.S. Pat. 4,148,835, Apr. 10, 1979b.

Chen, N. Y. and W. J. Reagan, U.S. Pat. 4,278,565, Jul. 14, 1981.

Chen, N. Y. and D. E. Walsh, U.S. Pat. 4,808,296, Feb. 28, 1989a.

Chen, N. Y. and D. E. Walsh, U.S. Pat. 4,877,581, Oct. 31, 1989b.

Chen, N. Y. and T. Y. Yan, Ind. Eng. Chem., Process Design Devel. *25*, 151 (1986).

Chen, N. Y., J. Maziuk, A. B. Schwartz, and P. B. Weisz, Oil Gas J. *66*(47) 154 (1968).

Chen, N. Y., R. L. Gorring, H. R. Ireland, and T. R. Stein, Oil Gas J. *75*(23), 165 (1977).

Chen, N. Y., W. E. Garwood, W. O. Haag, and A. B. Schwartz, "Shape Selective Hydrocarbon Catalysis Over Synthetic Zeolite ZSM-5," paper presented at Symp. Advan. Catal. Chem. I, Snowbird, Utah, Oct. 3–5, 1979a.

Chen, N. Y., W. W. Kaeding, and F. G. Dwyer, J. Am. Chem. Soc. *101*, 6783 (1979b).

Chen, N. Y., T. F. Degnan, and L. R. Koenig, Chemtech *16*, 506 (1986).

Chu, C. T.-W. and C. D. Chang, J. Catal. *86*, 297 (1984).

Chu, P. C. and G. H. Kuehl, Ind. Eng. Chem. Res. *26*, 365 (1987).

Chu, Y. F., D. O. Marler, and J. P. McWilliams, U.S. Pat. 4,885,426, Dec. 5, 1989.

Clerici, M. G., Appl. Catal. *66*, 249 (1991).

Coq, B., J. Pardillos, and F. Figueras, Appl. Catal. *62*, 281 (1990).

Corma, A., M. A. Camblor, P. Esteve, A. Martínez, and J. Perez-Pariente, J. Catal. *145*, 145 (1994).

Corma, A., P. Esteve, A. Martínez, and S. Valencia, J. Catal. *152*, 18 (1995).

Dadalla, A. M., T. C. Chan, and R. G. Anthony, Int. J. Chem. Kinet. *15*, 759 (1983).

Dartt, C. B. and M. E. Davis, Catal. Today, *19*, 151 (1995).

Das, J., Y. S. Bhat, and A. B. Halgeri, Catal. Lett. *23*, 161 (1994).

Dejaifve, P., J. C. Vedrine, and E. G. Derouane, J. Catal. *63*, 331 (1980).

Derouane, E. G., Stud. Surf. Sci. Catal. *20*, 221 (1985).

Derouane, E. G. and J. C. Vedrine, J. Mol. Catal. *8*, 479 (1980).

Dessau, R. M., J. Catal. *89*, 520 (1984).

Dessau, R. M. and R. B. LaPierre, J. Catal. *78*, 136 (1982).

Donnelly, S. P. and J. R. Green, "Recent MDDW Process Improvements," paper presented at the Symp. Jap. Petrol. Inst., Tokyo, Oct. 2–3, 1980.

Dwyer, F. G. and D. J. Klocke, U.S. Pat. 4,049,737, Sep. 20, 1977.

Eichler, K. and E. Leupold, U.S. Pat. 4,604,470, Aug. 5, 1986.

Eichler, K., E. Leupold, H. J. Arpe, and H. Baltes, German Pat. 3,420,707, Dec. 5, 1985.

Eichler, K., H. J. Arpe, and E. I. Leupold, European Pat. 175,228, Mar. 26, 1986.

Espinoza, R. F. and W. G. B. Mandersloot, J. Mol. Catal. *24*, 127 (1984).

Feitler, D., U.S. Pat. 4,612,387, Sep. 19, 1986.

Feitler, D., W. Schimming, and H. Wetstein, U.S. Pat. 4,675,410, Jun. 23, 1987.

Fowles, P. E. and T. Y. Yan, U.S. Pats. 4,524,227, 4,524,228, and 4,524,231, Jun. 18, 1985.

Fu., Z. H. and Y. Ono, Catal. Lett. *18*, 59 (1993); *22*, 277 (1993).

Galya, L. G., M. L. Occelli, J. T. Hsu, and D. C. Young, J. Mol. Catal. *32*, 391 (1985).

Garwood, W. E., Am. Chem. Soc. Symp. Ser. *218*, 383 (1983).

Garwood, W. E. and N. Y. Chen, "Octane Boosting Potential of Catalytic Processing of Reformate Over Shape Selective Zeolite," paper presented at the Am. Chem. Soc. Mtg., Houston, Mar. 23–28, 1980; Div. Petrol. Chem. Preprint, *25(1)*, 84 (1980).

Giannetto, G., R. Monque, and R. Galiasso, Catal. Rev.-Sci. Eng. *36(2)* 271 (1994).

Givens, E. N., C. J. Plank, and E. J. Rosinski, U.S. Pats. 4,079,095 and 4,079,096, Mar. 14, 1978.

Gorring, R. L., J. Catal. *31*, 13 (1973).

Gould, R. M., A. A. Avidan, J. L. Soto, C. D. Chang, and R. F. Socha, "Scale-Up of a Fluid-Bed Process for Production of Light Olefins from Methanol," paper presented at the AIChE Nat. Mtg., New Orleans, April 6–10, 1986.

Haag, W. O., Proc. 6th Int. Zeolite Conf., July 10–15, 1983, Reno, Nev., D. Olson and A. Basio, eds., Butterworths, Surrey, UK, p. 466, 1984.

Haag, W. O. and R. M. Dessau, Proc. 8th Int. Congr. Catal. *2*, 305 (1984).

Haag, W. O. and F. G. Dwyer, "Aromatics Processing with Intermediate Pore Size Zeolite Catalysts," paper presented at the AIChE 8th Nat. Mtg., Boston, Aug. 19, 1979.

Haag, W. O. and R. M. Lago, U.S. Pat. 4,374,296, Feb. 15, 1983a.

Haag, W. O. and R. M. Lago, U.S. Pat. 4,418,235, Nov. 29, 1983b.

Haag, W. O. and D. H. Olson, U.S. Pat. 4,117,026, Sep. 26, 1978.

Haag, W. O. and J. G. Santiesteban, U.S. Pat. 5,177,281, Jan. 5, 1993.

Haag, W. O., P. G. Rodewald, and P. B. Weisz, "Catalytic Production of Aromatics and Olefins from Plant Materials," paper presented at the Am. Chem. Soc. Nat. Mtg, San Francisco, August 24–29, 1980.

Haag, W. O., R. M. Lago, and P. B. Weisz, Chem. Soc. Faraday Disc. *72*, 317 (1981).

Haag, W. O., R. M. Lago, and P. G. Rodewald, J. Mol. Catal. *17*, 161 (1982).

Haag, W. O., R. M. Lago, and P. B. Weisz, Nature *309*, 589 (1984).

Hagen, G. P., U.S. Pat. 4,433,174, Feb. 21, 1984.

Hari Prasad Rao, P. R., P. Massiani, and D. Barthomeuf, Catal. Lett. *31*, 115 (1995).

Heinemann, H., Catal. Rev.-Sci. Eng. *15*, 53 (1977).

Hölderich, W., Proc. 7th Int. Zeolite Conf., Y. Murakami, A. Iijima and J. W. Ward, eds., Kodadansha/Elsevier, p. 827, 1986.

Hölderich, W. F., Stud. Surf. Sci. Catal. *58*, 631 (1991).

Hölderich, W., F. Merger, W. D. Mross, and R. Fischer, European Pat. 162,387, Nov. 27, 1985a.

Hölderich, W., F. Merger, W. D. Mross, and G. Fouquet, U.S. Pat. 4,560,822, Dec. 24, 1985b.

Hölderich, W., F. Merger, and W. D. Mross, U.S. Pat. 4,694,107, Sep. 17, 1987.

Hölderich, W., M. Hesse, and F. Näumann, Agew. Chem. Int. Ed. Engl. *27*, 226 (1988).

Hölderich, W., N. Goetz, and G. Fouquet, U.S. Pat. 5,079,367, Jan. 7, 1992.

Huang, T. J. and W. O. Haag, U.S. Pat. 4,339,606, Jul. 13, 1982.

Huang, T. J., C. M. Sorensen, and P. Varghese, U.S. Pat. 4,906787, Mar. 6, 1990.

Hutchings, G. N., P. Johnston, D. F. Lee, A. Warwick, C. D. Williams, and M. Wilkinson, J. Catal. *147*, 177 (1994).

Ihara Chem. European Pat. 154236, Sep. 11, 1985a.

Ihara Chem. Japan Pat. 60-048943, Mar. 16, 1985b.

Innes, R. A., S. I. Zones, and G. J. Nacamuli, U.S. Pat. 5,081,323, Jan. 14, 1992.

Inui, T., T. Ishihara, and Y. Takegami, J. Chem. Soc., Chem. Commun. *936* (1981).

Ishida, H. and H. Nakajima, U.S. Pat. 4,613,717, Sep. 23, 1986.

Jacobs, P. A., J. B. Uytterhoeven., M. Steyns, G. Froment, and J. Weitkamp, Proc. 5th Int. Conf. Zeolites, L. V. Rees, ed., Heyden, London, p. 607, 1980.

Jacobs, P. A., J. A. Martens, J. Weitkamp, and H. K. Beyer, Chem. Soc. Faraday Disc. *72*, 353 (1981).

Kaeding, W. W., J. Catal. *120*, 409 (1989).

Kaeding, W. W., J. Catal. *95*, 512 (1985).

Kaeding, W. W. and S. A. Butter, J. Catal. *61*, 155 (1980).

Kaeding, W. W. and L. B. Young, U.S. Pat. 4,094,921, Jun. 13, 1978.

Kaeding, W. W., M. M. Wu, L. B. Young, and G. T. Burress, U.S. Pat. 4,197,413, Apr. 8, 1980.

Kaeding, W. W., C. Chu, L. B. Young, B. Weinstein, and S. A. Butter, J. Catal. *67*, 159 (1981a).

Kaeding, W. W., C. Chu, L. B. Young, and S. A. Butter, J. Catal. *69*, 392 (1981b).

Kaeding, W. W., L. B. Young, and A. G. Prapas, Chemtech *12*, 556 (1982).

Kaiser, S. W., U.S. Pat. 4,499,327, Feb. 12, 1985; U.S. Pat. 4,524,234, Jun. 18, 1985.

Karge, H. G., J. Ladebeck, Z. Sarbak, and K. Hatada, Zeolites *2*, 94 (1982).

Karge, H. G., Y. Wada, J. Weitkamp, S. Ernst, U. Girrbach, and H. K. Beyer, Stud. Surf. Sci. Catal. *19*, 101 (1984).

Keim, K., R. Kiauk, and E. Meisenburg, U.S. Pat. 4,283,571, Aug. 11, 1981.

Kitagawa, H., Y. Sendoda, and Y. Ono, J. Catal. *101*, 12 (1986).

Kresge, C. T., J. P. McWilliams, J. C. Vartuli, and M. P. Nicoletti, U.S. Pat. 4,547,605, Oct. 15, 1985.

Kunchal, S. K., F. G. Dwyer, and C. P. Ambler, "Isofin—A New Process for Olefin Isomerization," paper presented at the 1993 NPRA. Annual Mtg., San Antonio, TX, (AM-93-45), March 21–23, 1993.

Kwak, B. S., W. M. H. Sachtler, and W. O. Haag, J. Catal. *149*, 465 (1994).

Kwok, T. J. and K. Jayasuriya, J. Org. Chem. *59*, 4939 (1994).

Landis, P. S. and P. B. Venuto, J. Catal. *6*, 245 (1966).

LaPierre, R. B., R. D. Partridge, N. Y. Chen, and S. S. Wong, U.S. Pat. 4,419,220, Dec. 6, 1983.

LaPierre, R. B., R. D. Partridge, N. Y. Chen, and S. S. Wong, U.S. Pat. 4,501,926, Feb. 26, 1985.

Le Blanc, H., L. Puppe, and K. Wedemeyer, U.S. Pat. 4,628,097, Dec. 9, 1986.

Linstid, H. C. and G. S. Koermer, U.S. Pat. 4,537,995, Aug. 27, 1985.

Litterer, H., German Pat. 3,334,673, Apr. 11, 1985.

Long, G. N., R. J. Pellet, and J. A. Rabo, U.S. Pat. 4,528,414, Jul. 9, 1985.

Martens, J. A., "Mechanism of Isomerization and Hydrocracking of Long-Chain Paraffins on Large Pore and Shape Selective Zeolites," Ph.D. thesis, Catholic University Leuven, Belgium, Dec., 1985.

Miller, S. J., U.S. Pat. 4,859,311, Aug. 22, 1989.

Miller, S. J., "New Molecular Sieve Process for Lube Dewaxing by Wax Isomerization," paper presented at the Symp. on New Catalytic Materials Utilizing Molecular Sieves, Div. of Petrol. Chem., Am. Chem. Soc. Chicage Mtg. Aug. 22–27, 1993.

Minachev, Kh. M., O. K. Shchudina, M. A. Markov, and R. V. Dimitriev, Neftekhimiya *8*, 37 (1968).

Minachev, Kh. M., V. I. Garanin, T. A. Isakova, V. V. Kharlamov, and V. Bogomolov, Advan. Chem. Ser. *102*, 441 (1971).

Minachev, Kh. M., V. I. Garanin, V. V. Kharlamov, and T. A. Isakova, Kinet. Katal. *13*. 1101 (1972).

Nagy, J. B., J. Mol. Catal. *5*, 393 (1979).

Nakamura, T., K. Shimoda, and K. Yasuda, Chem. Lett. 1881–1882 (1992).

Neves, I., F. Jayat, P. Magnoux, G. Pérot, F. R. Ribeiro, M. Gubelmann, and M. Guisnet, J. Mol. Catal. *93*, 169 (1994).

Notari, B., Stud. Surf. Sci. Catal. *37*, 413 (1987).

Olson, D. H. and W. O. Haag, Am. Chem. Soc. Symp. Ser. *248*, 275 (1984).

Olson, D. H., G. T. Kokotailo, S. L. Lawton, and W. H. Meier, J. Phys. Chem. *85*, 2238 (1981).

Ono, Y. and T. Mori, J. Chem. Soc. Faraday Trans. *177*, 2209 (1981).

Onodera, T., T. Sakai, Y. Yamasaki, and K. Sumitani, U.S. Pat. 4,320,242, Mar. 16, 1982.

Oudejans, J. C., F. J. van der Gaag, and H. van Bekkum, Proc. 6th Int. Zeolite Conf., July 10–15, 1983, Reno, Nevada, D. Olson and A. Bisio, eds., Butterworths, Surry, UK, p. 536, 1984.

Parikh, P. A., N. Subrahmanyam, Y. S. Bhat, and A. B. Halgeri, J. Mol. Catal. *88*, 85 (1994).

Pecci, G. and T. Floris, Hydrocarbon Procss. *56*, 98 (1977).

Pellet, R. J., J. A. Rabo, G. N. Long, and P. K. Coughlin, "Reactions of C_X Aromatics Catalyzed by Aluminophosphate Based Molecular Sieves," paper presented at the 10th N. Am. Mtg. Catal. Soc., paper No. B-2, San Diego, May 17–22, 1987.

Pellet, R. J., P. K. Coughlin, E. S. Shamshoum, and J. A. Rabo, Am. Chem. Soc. Symp. Ser. *368*, 512 (1988).

Perego, G., G. Bellussi, C. Corno, M. Taramasso, F. Buonomo, and A. Esposito, Stud. Surf. Sci. Catal. *28*, 129 (1986).

Plank, C. J., E. J. Rosinski, and E. N. Givens, U.S. Pat. 4,157,293, Jun. 5, 1979.

Pouilloux, Y., N. S. Gnep, P. Magnoux, and G. Perot, J. Mol. Catal. *40*, 231 (1987).

Price, G. L. and V. Kanazirev, J. Catal. *126*, 267 (1990).

Quann, R. J., L. A. Green, S. A. Tabak, and F. J. Krambeck, "Chemistry of Olefin Oligomerization over ZSM-5 Catalyst," paper presented at the AIChE Nat. Mtg., New Orleans, April 6–10, 1986.

Ratnasamy, P., G. P. Babu, A. J. Chandwadkar, and S. B. Kulkarni, Zeolites, *6*, 98 (1986).

Ratnasamy, P., R. N. Bhat, S. K. Pokhriyal, S. G. Hegde, and R. Kumar, J. Catal. *119*, 65 (1989).

Reddy, J. S., S. Sivasanker, and P. Ratnasamy, J. Mol. Catal. *70*, 335 (1991).

Rodewald, P. G., U.S. Pat. 4,100,219, Jul. 11, 1978; U.S. Pat. 4,145,315, Mar. 20, 1979.

Sachtler, J. W., R. J. Lawson, and S. L. Lambert, U.S. Pat. 5,081,084, Jan. 14, 1992.

Sano, T., O. Kiyomi, H. Hagiwara, H. Takaya, H. Shoji, and K. Matsuzaki, J. Mol. Catal. *40*, 113 (1987).

Sato, H., U.S. Pat. 4,465,852, Aug. 14, 1984.

Sato, H., N. Ishii, and S. Nakamura, U.S. Pat. 4,499,321, Feb. 12, 1985.

Sato, H., S. Kawakami, H. Dohi, and K. Endo, U.S. Pat. 4,568,793, Feb. 4, 1986.

Sato, K., J. Dakka, and R. A. Sheldon, J. Chem. Soc., Chem. Commun. 1887 (1994).

Shen, Jian-Ping, J. Ma, Da-Zhen Jiang, and En-Ze Min, Catal. Lett. *26*, 291 (1994).

Simon, M. W., S. L. Suib and C.-I. O'Young, J. Catal. *147*, 484 (1994).

Singh, B. B. and R. G. Anthony, Prepr. Can. Symp. Catal. *6*, 113 (1979).

Singh, B. B., F. N. Lin, and R. G. Anthony, Chem. Eng. Commun. *4*, 749 (1980).

Singh, A. P., S. B. Kumar, A. Paul, and A. Raj. J. Catal. *147*, 360 (1994).

Smith, K. W., W. C. Starr, and N. Y. Chen, Oil Gas J. *78(21)*, 75 (1980).

Song, C. and K. Moffatt, Am. Chem. Soc. Mtg. Chicago, Aug. 22–27, 1993; Div. Petrol. Chem. Preprints *38*, 779 (1993).

Stock, L. M. and H. C. Brown, J. Am. Chem. Soc. *81*, 3323 (1959).

Sugi, Y. and M. Toba, Catal. Today *19*, 187 (1994).

Sugi, Y., T. Matsuzaki, Y. Hanaoka, Y. Kubota, J.-H. Kim, X. Tu, and M. Matsumoto, Catal. Lett. *26*, 181 (1994).

Sullivan, R. F., C. J. Egan, G. E. Langlois, and R. P. Sieg, J. Am. Chem. Soc. *83*, 1156 (1961).

Tabak, S. A., F. J. Krambeck, and W. E. Garwood, "Conversion of Propylene and Butylene over ZSM-5 Catalyst," paper presented at the AIChE Mtg., San Francisco, Nov. 25–30, 1984; AIChE J. *32* 1526 (1986).

Taramasso, M., G. Perego, and B. Notari, U.S. Pat. 4,410,501, Oct. 18, 1983.

Tatsumi, T., M. Nakamura, K. Yuasa, and H.-O. Tominaga, Catal. Lett. *10*, 259 (1991).

Toray Industries, Japan Pat. 60-042340, Mar. 6, 1985.

Toray Industries, Japan Pat. 61-050933, Mar. 13, 1986.

van den Berg, J. P., J. P. Wolthuizen, and J. H. C. van Hooff, Proc. 5th Int. Conf. Zeolites, L. V. Rees, ed., p. 649, Heyden, London, 1980.

van den Berg, J. P., J. P. Wolthuizen, A. D. H. Clague, G. Hays, R. Huis, and J. H. C. van Hooff, J. Catal. *80*, 130 (1983); J. Catal. *80*, 139 (1983).

van den Berg, J. P., K. H. W. Röbschläger, and I. E. Maxwell, "The Shell Poly-Gasoline and Kero (SPGK) Process," paper presented at the 11th North Am. Catal. Soc. Zeolite Symp., Dearborn, May 7–11, 1989.

van der Gaag, F. J., F. Louter, J. C. Oudejans, and H. van Bekkum, Appl. Catal. *26*, 191 (1986).

van Hardeveld, R., T. J. van de Mond, and F. H. Vandenbooren, U.S. Pat. 4,625,061, Nov. 25, 1986.

Venuto, P. B., J. Org. Chem. *32*, 1272 (1967).

Venuto, P. B., Microporous Mat. *2*, 297 (1994).

Wang, I., Tsai, and S.-T. Huang, Ind. Eng. Chem. Res. *29*, 2005 (1990).

Wei, J., J. Catal. *76*, 433 (1982).

Weigert, F. J., J. Org. Chem. *52*, 3296 (1987).

Weisz, P. B., W. O. Haag, and P. G. Rodewald, Science *206*, 57 (1979).

Weitkamp, J., P. A. Jacobs, and J. A. Martens, Appl. Catal. *8*, 123 (1983).

Wise, J. J., U.S. Pat. 3,251,897, May 17, 1966.

Wu, M. M. and W. W. Kaeding, J. Catal. *88*, 478 (1984).

Wunder, F. A. and E. I. Leupold, Angew. Chem. Int. Ed. Engl. *19(2)* 126 (1980).

Yashima, T. and Komatsu, Y., U.S. pat. 5,312,915, May 17, 1994.

Young, L. B., U.S. Pat. 4,301,317, Nov. 17, 1981a.

Young, L. B., U.S. Pat. 4,283,573, Aug. 11, 1981b.

Young, L. B., U.S. Pat. 4,365,084, Dec. 21, 1982.

Young, L. B., U.S. Pat. 4,371,714, Feb. 1, 1983.

Young, L. B., S. A. Butter, and W. W. Kaeding, J. Catal. *76*, 418 (1982).

Yurchak, S., J. E. Child and J. H. Beech, "Mobil Olefins Conversion to Transportation Fuels," paper presented at the 1990 NPRA. Annual Mtg., San Antonio, TX, (AM-90-37), March 25–27, 1990.

5

Applications in Petroleum Processing

I. GOALS OF IMPROVEMENT FOR ENVIRONMENTALLY CLEAN FUELS

Currently in the United States, transportation fuels constitute more than two-thirds of refined oil. Almost 35% of the crude oil refined per day passes through catalytic cracking units and more than 70 wt % of the cracked products end up as transportation fuels (Weekman and Chen, 1991). Ground level ozone, formed by photochemical reactions of nitrogen oxides and hydrocarbons in the air, is one of the major air pollutants of concern (Bell et al., 1995). Concerns about ozone, acid rain, and global warming have prompted a review of transportation fuels in the United States by both the government and environmental groups.

The current understanding of how individual gasoline components affect various environmental issues is very limited. However, aromatics—notably benzene, high vapor pressure hydrocarbons such as n-butane, and reactive hydrocarbons such as olefins and sulfur compounds—are now considered harmful if released to the environment. They can enter the atmosphere through spills and vaporization losses or as the result of incomplete combustion.

Fluid catalytic cracking (FCC) is the principal refinery upgrading process because of its ability to upgrade low-valued high-boiling feedstocks to high-valued transportation fuels. In addition to process improvements, major advances in catalysts have been made in catalytic cracking over the last 50 years.

Naphtha reforming provides a major high-octane source for the gasoline pool and also provides the hydrogen needed to balance the stoichiometric requirement of the rest of the product slate. The pros and cons of putting aromatics in gasoline remain to be resolved. Environmental pressure to provide hydrogen-rich transportation fuels and reduce aromatics, particularly benzene, in gasoline will force a change in the reforming process. A process to isomerize C_6 fraction instead of reforming may be favored. Other processes such as Mobil's M-Forming process, which will reduce the benzene content in reformates may also become commercially viable if the use of LPG (propane) as a alternative clean transportation fuel becomes attractive.

The increase in the production of high-octane isoparaffins, ethers, and alcohols will be constrained by the supply of light olefins. Thus, alternative FCC technology, including the development of catalysts which promote the selective production of light olefins, can have an enormous impact on the production of environmentally acceptable gasolines (Johnson and Avidan, 1993).

A variety of olefin upgrading processes could also provide a ready source of clean-burning premium quality gasoline and distillate transportation fuels (Yurchak et al., 1990; van den Berg, 1991; Harandi et al., 1991) without the need to alkylate isobutane. Olefin-based transportation fuels not only meet the hydrogen-rich requirement, but are also free of sulfur, a contaminant in all of today's fuels.

Current gasoline is predominantly a blend of hydrocarbons, including *n*-butane, isopentane and isomeric mixtures of 6- to 11-carbon paraffins, olefins, naphthenes, and aromatics formulated to meet the performance properties, such as volatility and octane rating, required by the carburated and fuel-injected internal combustion engines.

The diesel fuel is a similar blend of hydrocarbons but of higher molecular weight (up to 16 carbons) formulated to meet the handling, combustion, and cold flow properties, such as flash point, cetane index, and cloud point, required by diesel engines operating year round.

The production of transportation fuels involved a variety of catalytic processes, including light paraffin (butane through hexane) isomerization, isobutane alkylation, naphtha reforming, cracking, and hydrogenative processing such as desulfurization, hydrotreating, and hydrocracking, among others. Shape selective catalysis has made a number of inroads for improving the selectivity in producing high-octane gasoline blending components and clean burning distillates.

II. CATALYTIC CRACKING OF GAS OILS

The need to increase production of light olefins from FCC and reduce sulfur in FCC products provides an incentive to couple FCC with feed hydrotreating. Catalytic hydrotreating also satisfies the stoichiometric hydrogen to carbon balance and helps to reduce excessive catalyst deactivation due to coke deposition. With the need to invest in additional gas production facilities comes the incentive to develop advanced gas separation technology (Johnson and Avidan, 1993).

The addition of a small concentration of ZSM-5 to the fluid catalytic cracking reactor or the incorporation of ZSM-5 in the conventional catalyst boosts the octane rating of the gasoline from the catalytic cracking units by 1 to 3 numbers. ZSM-5 increases research and motor octane number by selectively upgrading low octane components in the gasoline boiling range to higher-octane, lower molecular weight compounds. ZSM-5 accomplishes this selectively, without increasing methane, ethane, hydrogen, or coke (Donnelly et al., 1987; Dwyer et al., 1987). Principal gas components are propene, butenes, and isobutane, with the increase in propene being typically twice that of isobutene (Madon, 1991). The isobutane can be fed to an alkylation unit to produce additional high-octane gasoline, raising the octane pool of the refinery (Yanik et al., 1985).

The process has been tested commercially in both TCC and FCC units. Results of a commercial test (Anderson et al., 1984) conducted in a 15,000 bar-

Table 5.1 Commercial Test of ZSM-5–Containing Cracking Catalysts in TCC

	Base catalyst	Catalyst containing ZSM-5	
Time on stream, days	0	72	106
Conversion, vol %	53.0	53.0	53.0
C_5^+ gasoline, vol %	42.3	40.9	41.0
Butenes, vol %	3.8	4.6	4.2
Propene, vol %	3.7	4.7	4.2
Light fuel oil, vol %	29.9	29.1	26.4
Research octane number R + 0	86.0	90.2	91.2
Motor octane number M + 0	77.4	79.2	79.5
Potential alkylate, vol %	12.7	15.7	14.2
Total gasoline, vol %	55.0	56.6	55.2

Source: Anderson et al. (1984).

Table 5.2 Commercial Test of ZSM-5–Containing Cracking Catalyst in FCC

	Base catalyst	Catalyst containing ZSM-5
Time on stream, days	0	37
Conversion, vol %	73.2	73.2
C_5^+ gasoline, vol %	53.9	51.6
Butenes, vol %	7.9	9.0
Propene, vol %	6.2	7.0
Light fuel oil, vol %	26.5	26.5
Coke, wt %	6.7	6.7
Research octane number R + 0	87.3	89.0
Motor octane number M + 0	78.1	78.7
Potential alkylate, vol %	24.9	28.3
Total gasoline, vol %	78.8	79.9

Source: Yanik et al. (1985).

rels per day (B/D) TCC unit are shown in Table 5.1. In this test, the regular makeup catalyst (about 2 tons/day) was replaced by the new catalyst containing ZSM-5 for 93 days followed by the regular makeup catalyst. It was noted that even at 34 days after the addition of the ZSM-5–containing catalyst was stopped, the catalyst inventory (about 350 tons) in the unit continued to show octane-enhancing properties. Similar results (Table 5.2) have been obtained commercially in FCC units (Yanik et al., 1985).

Since then, at least 45 FCC and TCC units, ranging in size from 6,000 to 90,000 B/D, have used ZSM-5 additives. Schipper et al. (1988) reported significant increases in propene, and butenes were noted with ~1.5 wt % ZSM-5 in the inventory.

Johnson and Avidan (1993) studied the effect of adding more ZSM-5 in FCC to promote the production of light olefins. This can have an enormous impact on the production of environmentally acceptable gasolines (Johnson and Avidan, 1993). By adding 10 to 20% ZSM-5 to the catalyst inventory, a significant yield of light olefins was obtained (Table 5.3).

Buchanan (1991) examined the cracking characteristics of model compounds over steamed deactivated ZSM-5 and over admixtures of steamed ZSM-5 and REY or USY under simulated FCC conditions. Reaction temperatures and residence times were selected to approximate typical FCC conditions. The paraffins showed little reactivity; conversions were less than 1%. Cracking of the olefins ranged from 26% for hexene to 80% for decene. Hexene cracking experiments over sequential beds of ZSM-5 and Y zeolite base

Table 5.3 Maximum FCC Light Olefins Yields

	Case A	Case B
Riser temperature, °C	549	549
ZSM-5, % inventory	10	20
Conversion, vol %	88.3	89.0
Yield, vol %		
Gasoline	52.8	47.1
Isobutane	7.4	7.8
n-Butane	2.0	2.0
Butenes	16.3	18.6
Propane	4.3	4.3
Propene	15.6	22.0
C_2 *and lighter, wt %*	5.7	5.7
Total cycle oil	11.7	11.0
Coke, wt %	7.8	7.8
RON	96.7	97.3
MON	84.4	84.6

Source: Johnson and Avidan (1993).

catalysts demonstrated that the removal of olefins by ZSM-5 reduced the secondary hydrogen transfer reactions which formed the paraffins and aromatics over REY or USY catalysts. Madon (1991) reached similar conclusions in his pilot scale riser studies.

Chen et al. (1988) studied catalytic cracking with a mixture of faujasite and zeolite Beta. A waxy gas oil (properties shown in Table 5.4) was used as the feedstock. Under standard TCC operating conditions, the catalyst mixture yielded a higher octane gasoline (Figure 5.1), higher yields of C_3 and C_4 olefins (Figure 5.2), and lower pour point distillate fuels. Figure 5.3 shows the result on the vacuum distillate bottoms.

III. OCTANE BOOSTING

A. Reforming of Naphthas

Reforming of naphtha to high-octane gasoline uses Pt-Al$_2$O$_3$ (or bimetallic) catalyst under hydrogen pressure to carry out a number of reactions, including converting 6-ring and 5-ring naphthenes to aromatics via dehydrogenation and dehydroisomerization and converting paraffins to higher-octane isoparaffins

Table 5.4 Waxy Gas Oil Properties

API gravity	33.8
Pour point, °C	40.6
Hydrogen, wt %	13.67
Sulfur, wt %	0.15
Nitrogen, ppm	180
Nickel, ppm	0.14
Vanadium, ppm	0.10
Molecular weight	313
Paraffins, wt %	62.5
Naphthenes, wt %	12.3
Aromatics, wt %	24.7
IBP, °C	205
End point, °C	485

Source: Chen et al. (1988).

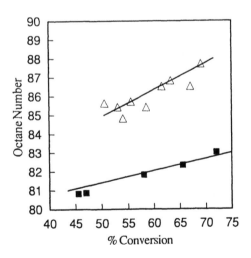

Figure 5.1 Octane number of gasoline versus conversion: (Δ) REY + zeolite Beta; (■) REY only. (From Chen et al., 1988.)

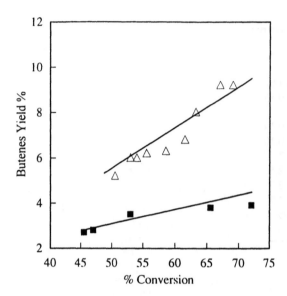

Figure 5.2 Butenes yield (vol %) versus conversion. (From Chen et al., 1988.)

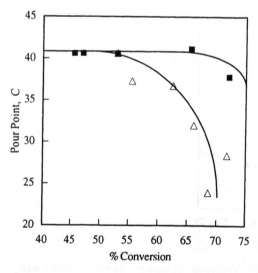

Figure 5.3 Pour point of distillate versus conversion. (From Chen et al., 1988.)

and aromatics via hydroisomerization and dehydrocyclization. However, two limitations are inherent in this "reforming" technology. First, the product always contains an essentially equilibrium distribution of paraffin isomers and hence a significant amount of low-octane normal paraffins remains in the product, limiting the product octane. Second, not only is the paraffin dehydrocyclization efficiency for C_7^- paraffins quite low (Ramage et al., 1987) (Figure 5.4), there is also a significant volume contraction when paraffins are converted to aromatics, lowering the volumetric product yield.

Thus, the reforming of light naphthas produces a low yield of reformate. In the case of full range naphthas, high severity reforming, above 98 R + 0, is generally achieved by concentrating the aromatics and hydrocracking the low-octane component to gases; hence, this route also produces a low liquid yield.

The addition of a small amount of low-acidity ZSM-5 to the reforming catalyst (Plank et al., 1979, 1981) selectively cracks the low-octane paraffinic molecules to propane and butanes and increases the octane boosting potential

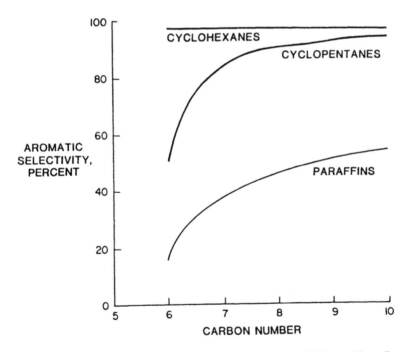

Figure 5.4 Component reforming efficiencies at 480°C and 15 atm. (From Ramage et al., 1987.)

Table 5.5   Effect of ZSM-5 on Reforming Catalyst Performance

Feed: C_6-140°C midcontinent naphtha

Operating conditions:
Pressure: 14.6 atm, LHSV: 1.7,
hydrogen/hydrocarbon: 9.6 mol/mol

Catalyst	Base case	Base catalyst + 1 wt % ZSM-5	
Temperature, °C	485	485	465
C_5^+ Octane	94.5	101.8	96.4
C_5^+ Yield, vol %	83.4	69.7	74.4
Gas yields, wt %			
methane, ethane	1.8	3.0	1.6
propane	2.4	9.1	7.7
butanes	2.9	9.8	9.1

Source: Plank et al. (1979).

of the reforming catalyst. For example, the data in Table 5.5 show that under identical operating conditions, the reforming catalyst containing 1 wt % of ZSM-5 increased the C_5^+ octane by 7.3 numbers over the base case. The comparison of these two catalysts at the same octane severity shows that the ZSM-5 containing reforming catalyst is about 20°C more active than the base case. The gain in propane and butanes also leads to a higher purity recycle gas, a key factor for stable operation.

B. Postreforming of Reformates

Selectoforming

In the mid-1960s, a shape selective postreforming process, "Selectoforming," was introduced (Chen et al., 1968). Selectoforming was the first commercial molecular shape selective catalytic process. It was designed to upgrade the octane rating of the reformate and produce LPG (primarily propane) as the major byproduct. It is operated under hydrogen pressure and uses erionite, an 8-membered ring small pore zeolite, as the catalyst (see Chapter 2).

Based on the principle of size exclusion, the Selectoforming catalyst selectively converts the low octane *n*-paraffins in a reformate to propane. In addition to size exclusion of the feed molecules, the selective production of propane instead of isobutane as the principal product of acid cracking represents another example of the size exclusion principle as applied to the product molecules.

Table 5.6 Chemical Composition of Natural Erionite and Its H-Form

As received, wt %		Cation distribution equiv./Al		H-form cation distribution equiv./Al
SiO$_2$	65.9			
Al$_2$O$_3$	15.0			
K$_2$O	3.5	K	0.25	0.25
CaO	3.0	Ca	0.36	0.06
Na$_2$O	6.7	Na	0.73	0.01
MgO	0.7	Mg	0.14	0.05
Fe$_2$O$_3$	3.8	Fe	0.32	H+ 0.63
Others	1.2		1.80	1.00
	100.0	SiO$_2$/Al$_2$O$_3$ = 7.5		SiO$_2$/Al$_2$O$_3$ = 7.4

Source: Chen and Garwood (1973a).

To assure catalyst stability, a weak metal function, such as the sulfides of nickel, was incorporated into the catalyst to saturate olefins and prevent coke formation, but not to hydrogenate the aromatics present in the feed (Chen and Garwood, 1968; Chen and Rosinski, 1971).

The acidity of the catalyst was generated by ion exchanging the zeolite with ammonium ions (Table 5.6). About 20–25% of the exchangeable sites are occupied by potassium ions, situated inside the small cancrinite cages. This residual potassium may be removed by additional calcination and ammonium exchange steps. However, at below 450°C, the maximum operating temperature of the process—the rate of migration of potassium which deactivates the catalyst—is very slow. Shown in Figure 5.5 is the commercial performance of a catalyst prepared without removing the residual potassium. The catalyst had a cycle life in excess of 1½ years.

Figure 5.6 shows a number of reactor configurations by which the Selectoforming process has been practiced commercially. Placing the catalyst in the bottom of the last reforming reactor is a simple and low cost approach, but it does not have the flexibility of using a separate reactor which can be operated at a different temperature and/or on a part-time basis when the market for propane is good.

The process generally operates at the reformer pressure. Reactor temperature may vary between 315°C and the temperature of the last reformer reactor, depending on the reactor configuration. Recycle gas containing 60 to 80 mol % hydrogen may be used and the gas-to-oil ratio is maintained at between 2 and 4.

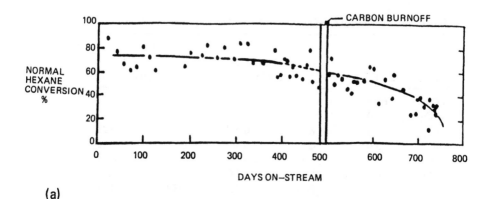

(a)

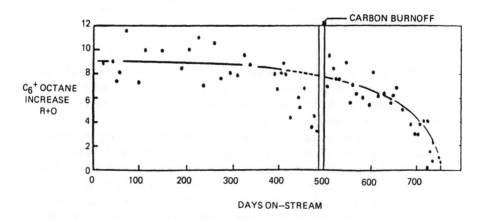

(b)

Figure 5.5 (a) Normal hexane conversion across selectoforming catalyst in commercial operation. (b) Research clear octane number increase across selectoforming catalyst in commercial operation. (From Roselius et al., 1973.)

The charge stocks can be unstabilized, debutanized, or depentanized reformates. Table 5.7 shows the yield shift after processing an unstabilized reformate. Case 1 is a situation where the reformer throughput is increased, and the low-octane intermediate product from the reformer is processed over the Selectoforming reactor to the same final octane level to co-produce propane. Case 2 shows the ability of the Selectoforming catalyst to boost the octane of the product beyond that of the reformer.

SELECTOFORMING IN THIRD REACTOR OF REFORMER TRAIN (HIGH TEMPERATURE)

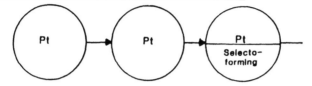

SELECTOFORMING IN THIRD REACTOR OF REFORMER TRAIN (LOW TEMPERATURE)

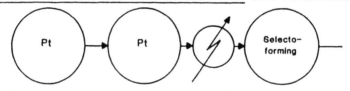

SELECTOFORMING IN SEPARATE REACTOR

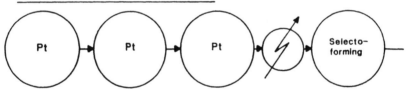

Figure 5.6 Catalyst fill configurations for reforming/Selectoforming process.

It is interesting to note that because the concentration of isoparaffins is higher in the Selectoformate than in that of conventional reformates at the same research octane, Selectoformate has a higher motor octane number, which is translated to higher road octane ratings (Chen et al., 1968; Burd and Maziuk, 1972).*

To produce propane, the Selectoforming process may be directly applied to light naphthas without reforming. Shown in Table 5.8 is the pilot plant result of Selectoforming a C_5-82°C light naphtha. The process selectively converts C_5^+ n-paraffins to propane and n-butane and raises the octane rating of the remaining liquid product. Although not in commercial practice, laboratory data

*Research octane number and a motor octane number are two standard laboratory measures of a fuel's ability to resist knock during combustion in a spark-ignited engine. These methods have been established by ASTM (1973) and are conducted in the laboratory by running the fuel in a single-cylinder test engine under prescribed operating conditions. The road octane number is an average of these two numbers (Benson, 1976).

Table 5.7   Selective Cracking for Propane and Selectoformate

	Case 1		Case 2	
Yields, wt %	Charge	Products	Charge	Products
C_1, C_2, H_2	7.0	8.7	13.6	16.6
C_3	5.3	15.3	10.8	15.5
C_4	5.9	6.3	10.8	8.9
i-C_5	6.4	7.6	8.7	8.0
n-C_5	6.1	1.6	7.0	3.9
C_6 + n-paraffins	8.9	1.8	2.6	1.3
C_6 + iso's + aromatics	60.4	58.7	46.5	45.8
Octane numbers	Reformate	Selectoformate	Reformate	Selectoformate
C_5^+, RON + 0	86.1	93.3	96.5	99.5
C_5^+, RON + 3	96.3	100.8	102.6	104.7
C_5', DON + 3[a]	93.0	98.5	100.2	103.9
C_6', RON + 0	87.3	94.3	102.9	104.7
C_6', RON + 3	97.1	101.3	106.3	107.6

[a]Distribution octane number: a measure of front-end octane quality.
Source: Chen et al. (1968).

Table 5.8   Selectoforming of a C_5-82°C Light Naphtha

Days on stream		25.6	2.6
Temperature, °C		377	424
LHSV		1.6	1.6
Octane, RON	67.4	76.5	83.1
Composition	Feed	Products	
Methane + ethane		1.5	3.9
Propane		18.2	25.7
Isobutane		0.5	1.3
n-Butane		6.0	8.5
Isopentane	22.2	22.2	22.2
n-Pentane	23.4	12.9	3.9
C_6^+	54.4	38.7	34.5

Source: Heck and Chen (1993).

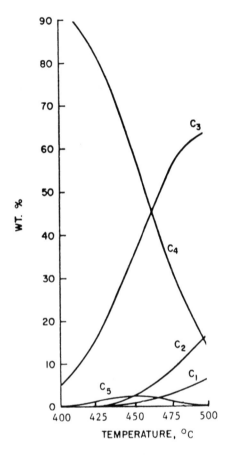

Figure 5.7 Product distribution from hydrocracking *n*-butane. (From Chen and Garwood, 1973a.)

indicate that under more severe operating conditions *n*-butane can be selectively converted to propane (Chen and Garwood, 1973a). Shown in Figure 5.7 is the product distribution from *n*-butane conversion over erionite at 21 atm and 1.5 LHSV over a temperature range of 400–510°C.

A commercial process converting *n*-butane and isobutane selectively to propane was developed by IFP (Fromager et al., 1977) using mordenite containing a non–noble metal hydrogenation function (Bernard et al., 1976). The range of operating conditions for this process is similar to those of Selectoforming.

Liers et al. (1993), using erionite containing reforming catalysts, demonstrated the advantage of increasing the erionite concentration in lowering the operating pressure from 25 to 15 atm.

To produce propane, the Selectoforming process may be directly applied to light naphthas without reforming. This process may become attractive again if a large demand for propane or LPG develops.

M-Forming

Another postreforming process using ZSM-5 as the catalyst, known as the M-Forming process (Chen, 1973; Heinemann, 1977) was developed in the early 1970s at Mobil. The catalyst performs two major functions: to selectively crack paraffins and to alkylate benzene and toluene present in the reformate with the olefinic portion of the cracked products. The rate of alkylation favors benzene and toluene over the higher alkylaromatics in the reformate (Chen and Garwood, 1978a; Garwood and Chen, 1980). These lighter alkylated products provide the additional liquid yield and retain the high octane rating of benzene and toluene, while heavy aromatics outside the gasoline boiling range are not produced to any significant extent due to shape selective constraints. The process improves liquid yield and reduces the concentration of benzene in the product.

The process may be operated with or without hydrogen circulation. However, in commercial practice as part of conventional reforming, it is convenient to integrate the M-Forming process into the reforming loop with continued hydrogen circulation.

As discussed in Chapter 3, unlike that for erionite, the rate of cracking of paraffins over ZSM-5 increases with the length of the molecules and decreases with the bulkiness of the molecules (See Figure 4.8). For octane boosting, this is a most desirable feature because the octane rating of paraffinic molecules increases in a similar fashion. Figure 5.8 shows the relationship between the relative cracking rate of C_5 to C_7 paraffins and their octane ratings.

To obtain yield improvement with M-Forming, it is important to properly interface the process with reforming. Generally speaking, this interfacing may be done on the basis of the octane-to-yield relationship. M-Forming is best interfaced at an octane severity beyond which reforming yield loss becomes excessive. This may depend on the composition and the boiling range of the naphtha, and the specific reforming process. On the other hand, M-Forming may indirectly influence the choice of the reforming process, e.g., a combination of ultra–low pressure reforming at moderate severities and M-Forming

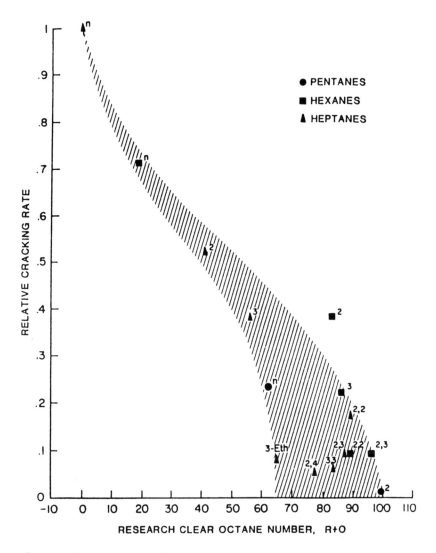

Figure 5.8 Octane rating versus cracking rate of paraffins. (From Chen et al., 1979.)

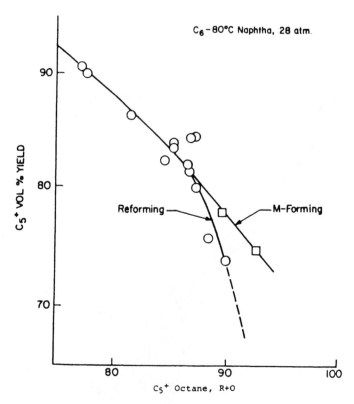

Figure 5.9   Reforming versus M-Forming. (From Chen et al., 1987.)

could achieve better liquid yield and cycle life than current reforming processes, which encounter catalyst stability problems at high reaction severities.

Figure 5.9 shows a typical yield/octane curve comparing regular reforming at 28 atm with an added M-Forming step (Bonacci and Patterson, 1981; Chen et al., 1987). In this case, the feed is a C_6, 80°C midcontinent light naphtha. The "interface" between reforming and M-Forming was made at 85 clear octane (C_5^+). It is noted that reforming beyond this "interface" point gave a lower liquid yield than M-Forming.

Table 5.9 shows detailed compositional changes at two different M-Forming severities (Chen et al., 1987). Aromatic alkylation is indicated by the disappearance of benzene and toluene and the appearance of C_8^+ aromatics.

For higher boiling naphthas, the "interface" is usually much higher, above 90 R+0, in order to achieve yield advantage over reforming.

Table 5.9 Compositional Change at Different M-Forming Severities

C$_5^+$, R+0	C$_6$-80°C mid-continent naphtha, 28 atm				
	Feed 84.5	Product 89.6		Product 92.7	
Aromatics					
B	18.5	16.9	−1.6	16.3	−2.2
T	23.4	22.5	−0.9	21.7	−1.7
X	0.6	1.1	+0.5	1.5	+0.9
C$_9$	0.2	2.9	+2.7	3.7	+3.5
C$_{10}$	0	2.0	+2.0	3.3	+3.3
C$_{11}$	0	0.3	+0.3	0.4	+0.4
Total	42.7	45.8	+3.0	46.8	+4.1
Paraffins					
C$_6$	30.9	26.2	−4.7	23.4	−7.5
C$_7$	12.9	9.7	−3.2	8.7	−4.2
C$_8$	0.2	0	−0.2	0.3	+0.1
C$_9$	0	0	0	0	0
Total	44.0	35.9	−8.1	32.4	−11.6
Naphthenes	1.4	1.6	+0.2	1.1	−0.3
Pentanes	8.0	8.6	+0.6	9.0	+1.0
Butanes and ligher	3.9	8.1	+4.2	10.7	+6.8

Source: Chen et al. (1987).

Shown in Figure 5.10 are the results obtained by interfacing M-Forming with reforming a full range C$_6$ to 165°C Persian Gulf naphtha at 76, 86, and 91 octane severities. It is noted that the octane of the M-formate increased at a nearly constant rate of about 1 octane for every 1.4 vol % loss of C$_5^+$ yield irrespective of the octane of the reformate. The yield loss by reforming for this naphtha is less than that of M-Forming up to about 94 R+0. Beyond 94 R+0, the yield loss by reforming becomes progressively greater than by M-Forming. Therefore, when the target octane exceeds 98 R+0, M-Forming clearly has a yield advantage over reforming alone.

Compared to conventional reformates at the same octane rating, the M-Forming products are low in aromatics and high in isoparaffins. Table 5.10 shows a typical example (Chen et al., 1987). Feedstock was a C$_6$ to 165°C Persian Gulf naphtha, the reforming was carried out at 18 atm over a Pt–Re bimetallic catalyst, and the product of M-Forming was made after reforming the naphtha to 84 clear octane and continued with M-Forming to 96.8 R+0.

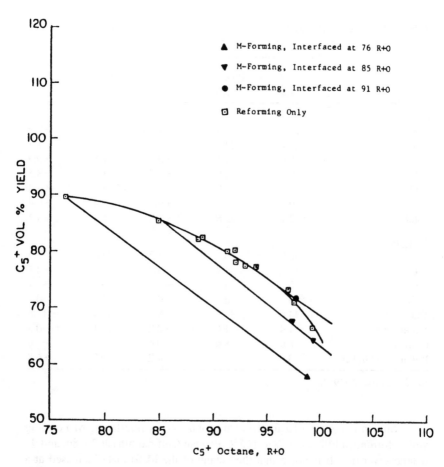

Figure 5.10 Reforming/M-Forming C$_6$-165°C Persian Gulf naphtha. (From Chen et al., 1987.)

Interfacing at such a low octane severity did not achieve a yield advantage for M-Forming; however, the product composition clearly demonstrated the reduction in aromatics compared to the same naphtha reformed to the same octane.

Environmental pressure to provide hydrogen-rich transportation fuels and reduce aromatics, particularly benzene, in gasoline will force a change in the reforming process. A process to isomerize C$_6$ fraction instead of reforming may be favored; other process such as benzene hydrosaturation may be considered. However, each of them has undesirable features such as loss of hydro-

Table 5.10 Comparison of M-Forming and Reforming of a C_6-165°C Persian Gulf Naphtha

	M-Forming product	Reforming product	
C_5^+ R + 0	96.8	96.7	
Aromatics, wt %			
benzene	5.5	5.5	0
toluene	15.7	16.3	−0.6
C_8	18.9	19.0	−0.1
C_9	11.0	17.5	−6.5
C_{10}	3.4	2.9	+0.5
C_{11}	1.3	0.1	+1.2
Subtotal	55.8	61.4	−5.6
Isopentanes	7.9	6.9	+1.0
n-Pentane	3.6	4.8	−1.2
Isohexanes	13.9	13.5	+0.4
n-Hexane	1.6	5.6	−4.0
Isoheptanes	8.0	5.3	+2.7
n-Heptane	0.1	1.5	−1.4
Isooctanes	4.3	0.9	+3.4
n-Octane	0	0.1	−0.1
Isononanes	1.3	0	+1.3
n-Nonane	0	0	0
Isodecanes	0.1	0	+0.1
n-Decane	0	0	0
Naphthenes	3.4	0	+3.4
Subtotal	44.2	38.6	+5.6
Total	100	100	

Source: Chen et al. (1987).

gen supply from the reformer and loss of octane rating. Mobil's M-Forming process can be adapted to coprocess reformate with the light olefins from the FCC unit (Evitt et al., 1992) by alkylating benzene to alkylaromatics. The process is known as the MBR process.

An example is given in Figure 5.11. In this example, a basic MBR unit processes 2000 B/D benzene-rich heart-cut reformate distilled out the debutanized reformate plus the olefinic FCC fuel gas. Benzene is reduced from an estimated 668 B/D in the reformate along with a substantial octane gain. The

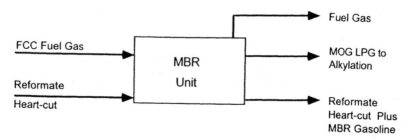

Figure 5.11 Benzene reduction in heart-cut reformate via MBR unit.

Effect on total refinery gasoline pool

	Base case	+MBR
Olefins, vol%	13.3	13.5
Benzene, vol %	1.8	0.9
Aromatics, vol %	29.2	29.4
Gasoline production, BPSD	50,000	50,300

(From Evitt et al., 1992.)

gasoline pool benzene content drops from 1.8 vol % to less than 0.9 vol %. Further reduction of the benzene content can be achieved by incorporation of a recycle.

In addition to the MBR process, benzene reduction can be achieved by first extracting the benzene from the reformate followed by Mobil/Badger Cumene Process or the Mobil/Badger Ethylbenzene Process. The MBR process is now licensed exclusively through Badger (Goelzer et al., 1993).

IV. LIGHT OLEFINS UPGRADING PROCESSES

The chemistry of olefin isomerization as discussed in Chapter 4 offers a plateful of unique processes and products. Among them is Isofin, a new process for olefin isomerization, jointly being developed by BP Oil, Mobil, and Kellogg (Kunchal et al., 1993), and Mobil's MOI process for olefin interconversion (Ram and Chin, 1993).

The Isofin process is designed to isomerize linear olefins to branched olefins. More specifically the conversion of *n*-butenes to isobutene and the conversion of *n*-pentenes to isoamylenes. These branched olefins are perfect feed-

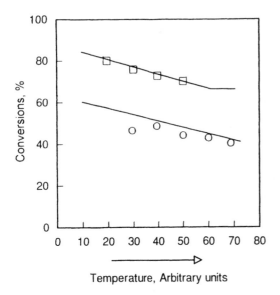

Figure 5.12 Skeletal isomerization of C_4 and C_5 olefins: (□) pentenes; (○) butenes. (From Kunchal et al., 1993.)

stocks for the production of oxygen-containing high-octane gasoline component by the synthesis of methyl-*tert*-butyl ether (MTBE) and *tert*-amyl-ether (TAME).

Since *n*- to iso-olefin isomerization is a thermodynamic equilibrium reaction, the iso/normal ratio is dependent on temperature. Figure 5.12 shows that the equilibrium percentage of isobutene varies from about 62% to about 42%, while isopentenes decrease from about 89% to about 71% over the same temperature range. Also shown in Figure 5.12 are the experimental data obtained with Mobil's catalyst. The temperature operating range is 350 to 425°C for butenes and 230 to 355°C for pentenes.

Mobil's MOI (Mobil Olefin Interconversion) process effectively converts heavy olefins over ZSM-5 to produce lighter olefins for potential clean fuel applications and petrochemical applications. The process operates at high space velocities combined with high temperatures and low pressures, which shift the equilibrium olefin distribution to lower carbon numbers. Figure 5.13 shows the effect of reaction temperature on the effluent olefin distribution. The process operates in a dense fluid-bed reactor/regenerator system very similar to that for the MOG process, as will be discussed in Section V. The process is licensed through Badger Technology Center of Raytheon Engineers & Constructors, Inc.

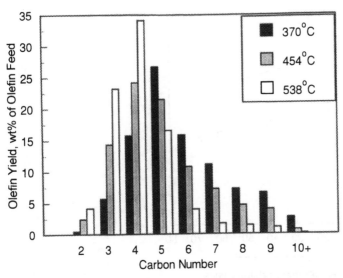

Figure 5.13   Effect of reaction temperature on the effluent olefin distribution. (From Ram and Chin, 1993.)

V.   CLEAN FUELS FROM LIGHT OLEFINS

Clean fuels may also be produced from light olefins. The oligomerization, transmutation/disproportionation, and aromatization of C_2 to C_{10} olefins, discussed in Chapter 4, form the basis of Mobil's Olefin-to-Gasoline and Distillate (MOGD) Process (Tabak, 1984; Tabak et al., 1984, 1986a) and the Mobil Olefins-to-Gasoline (MOG) process (Yurchak et al., 1990). The MOGD process is analogous to Mobil's Methanol-to-Gasoline (MTG) process. In the distillate mode, the products after hydrogenation are premium quality, low pour point jet fuel and distillate fuels. They are free of sulfur and aromatics. The process is operated under elevated pressures and low temperatures to produce iso-olefins; in the gasoline mode, the process is operated at higher reaction temperatures to increase aromatics production. Figure 5.14 shows that as the pressure is increased to above 23 atm, propene at 200°C is converted to 165°C+ olefins at better than 70% yield (Garwood, 1983). The MOG process operates at a higher reaction severity than MOGD to allow hydrogen transfer reactions to be significant, converting light olefins to high-octane gasoline (Yurchak et al., 1990).

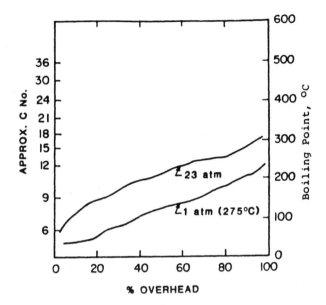

Figure 5.14 Effect of pressure on propene conversion at 0.4 WHSV and 200°C. (From Garwood, 1983.)

The selectivity of the catalyst plays a major role in producing these unique products. A schematic MOGD process flow diagram is shown in Figure 5.15. The design uses three reactors with interstage cooling and a condensed liquid recycle. While liquid recycle provides a heat sink for the control of reaction exotherm, it also affects the product quality by allowing the recycled material to be further reacted. In this design, both maximum gasoline and maximum diesel modes can be run by adjusting the reactor temperature and the boiling range of the recycle stream. Heavier recycle streams favor the production of distillates.

Table 5.11 presents the yield and quality of the MOGD products. As diesel, the paraffinic MOGD product is low in density but is an exceptionally good blending stock due to its low pour point and negligible sulfur content. Because of its paraffinic nature, the MOGD product makes excellent jet fuels, meeting or exceeding all commercial and military specifications. In particular, because of its low aromatic content, it is a superbly stable fuel as indicated by the JFTOT test.

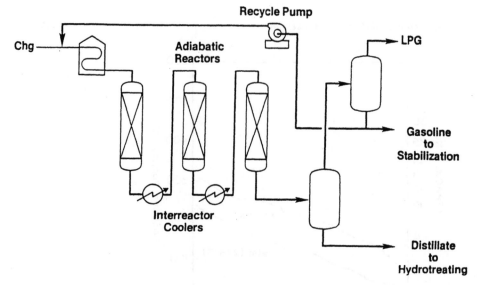

Figure 5.15 MOGD process flow, maximum distillate mode. (From Tabak et al., 1986a.)

Table 5.11 Yield and Quality of the MOGD Products

	MOGD product	Industry standard
Diesel fuel		
Specific gravity at 15°C	.79	.84–.88
Flash point, °C	60	52
Pour point, °C	≤50	−7
Cetane number	50	45
Sulfur, wt %	<.002	0.5 max
Viscosity, cs at 40°C	2.5	1.9–4.1
Boiling range, °C		
10%	203	249 max
50%	264	000 max
90%	324	338 max
Jet fuel		
Specific gravity	.79	.78–.8
Freezing point, °C	≤60	−40
Sulfur, wt %	<.002	0.3 max
Smoke point, mm	28	18 min
JETOT, °C	343	260
Hydrocarbon type		
FIA aromatics	4	25 max
olefins	1	5 max
paraffins	95	—

Source: Tabak et al. (1986b).

With the use of a C_3/C_4 mixed olefin stream from the refinery, the MOGD process was tested commercially in Mobil's Paulsboro Refinery in 1982.

The MOG process makes more coke than MOGD, and fluid-bed reactor systems with continuous catalyst regeneration is preferred than fixed-bed reactors. A schematic process flow diagram is shown in Figure 5.16. The fluid-bed system can tolerate wide swings in feed rate and feed quality and it easily overcomes upsets (Yurchak et al., 1990).

The mixture of branched olefins, paraffins, and alkylaromatics from the MOG reactions produces an excellent gasoline blend stock. Table 5.12 shows the composition and the properties of a typical gasoline product. It can be directly blended into the gasoline pool. The process has been successively demonstrated in a 4 B/D fluid-bed pilot plant over 8 months.

By operating the MOG unit at a higher severity, the olefin content drops from close to 60% to the light FCC gasoline down to about 13%. In this case

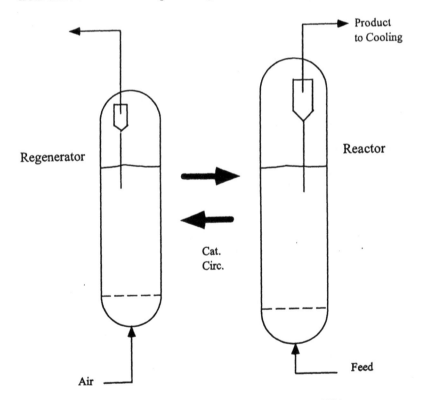

Figure 5.16 Fluid-bed MOG system. (From Yurchak et al., 1990.)

Table 5.12   Typical MOG C$_5{}^+$
Gasoline Composition and Properties

Specific gravity	0.73
Octane number	
R + 0	95
M + 0	82
Reid vapor pressure, psi	7.5
C$_5{}^+$ Composition, wt %	
Paraffins	31
Olefins	28
Naphthenes	8
Aromatics	33

Source: Yurchak et al. (1990).

the olefin content in the total gasoline pool drops from 13.3 vol % down to less than 5 vol % (Evitt et al., 1992). The MOG process is now licensed exclusively through Badger (Goelzer et al., 1993).

A similar process known as the Shell Poly-Gasoline and Kero (SPGK) process was developed by Shell Research in the Netherlands (van den Berg et al., 1989).

VI.   DEWAXING OF DISTILLATE FUELS AND LUBE BASESTOCKS

A.   Introduction

Shape selective cracking of straight chain and slightly branched chain paraffins over medium pore zeolites lends them to their application in the catalytic dewaxing of distillate fuels and lube basestocks. Figure 5.17 shows the boiling ranges and associated average carbon numbers for current distillate fuels (Weisz and Katzer, 1984). With current boiling point specifications on diesel fuel in the United States, use of pour point depressants and kerosene blending are required in the cold winter months to maintain acceptable fluidity. The cold flow properties (pour point, freezing point, cloud point, and cold filter plugging point) of distillate fuels are largely determined by the concentration of normal and slightly branched paraffins in the oil. Figure 5.18 shows the melting point of pure paraffins as a function of carbon number and boiling point for the normal paraffins. As discussed in Chapter 4, ZSM-5 and ZSM-11 having just the desired pore size to discriminate the waxy molecules from other bulkier mole-

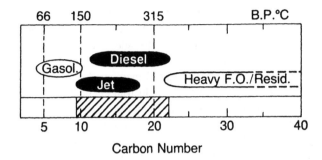

Carbon Number

Figure 5.17 Carbon number and boiling-point range of premium petroleum products. (From Wise et al., 1986.)

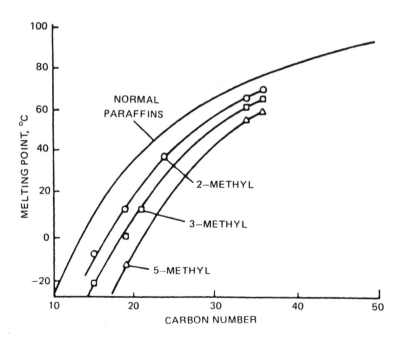

Figure 5.18 Melting points of normal and branched paraffins. (From Wise et al., 1986.)

cules, have been found to be effective dewaxing catalysts for a wide variety of feedstocks, ranging from jet fuels to resids. Other zeolites believed to have non-intersecting one-dimensional channel medium pore structures, such as Theta-1 (ZSM-22), ZSM-23, ZSM-35, ZSM-48, and SAPO-11, and a dual pore zeolite with interconnecting channels of elliptical 10- and 8-membered oxygen ring openings, such as synthetic ferrierite, have found industrial applications with the aid of a hydrogenation function, such as paraffin hydroisomerization over ZSM-22 and ZSM-23 (Chen et al., 1989) and over SAPO-11 (Miller, 1989). Mordenite, a dual pore zeolite with interconnecting channels of puckered 12- and 8-numbered oxygen ring openings, has also been found to be an effective lube dewaxing catalyst with the incorporation of a noble metal hydrogenative function (Bennett et al., 1975).

These dewaxing processes offer an opportunity to uncouple the historical relationship between cold flow properties and end point of distillate fuels and to increase the amount of high-value distillates with acceptable pour points producible from a given crude if the historical end point specification of the product is relaxed.

Catalytic dewaxing processes have also been developed to replace the expensive noncatalytic solvent dewaxing processes for lube oil processing. Catalytic dewaxing does not have the temperature limitation of a solvent dewaxing process. Thus, it creates a new class of super–low freeze point of pour point products for low temperature applications. The combination of commercial lube hydrotreating/hydrocracking processes and catalytic dewaxing made possible the development of an all-catalytic route to lubes.

B. Mobil Distillate Dewaxing Process

By the selective removal of waxy molecules from the oil using the medium pore zeolites, the process improves the fluidity of the oil at low temperatures. This is the basis of the Mobil Distillate Dewaxing (MDDW) process, first publicly announced in 1977 (Chen et al., 1977; Meisel et al., 1977). The process was commercialized in 1974. There are now more than 27 licensed units representing more than 160,000 barrels per day of installed capacity worldwide, including in the United States, Canada, Italy, Germany, Indonesia, and China (Smith and Bortz, 1990; Buyan et al. 1995).

MDDW is a fixed-bed process. Figure 5.19 shows a schematic flow diagram of an MDDW unit. The process is characterized by its long operating cycles (6 months to 1 year) between regenerations. Typical operating conditions for the MDDW process are 20 to 55 atm reactor pressure, 260–430°C reactor temperature, and 40–70 m^3/B of hydrogen circulation (Chen et al., 1977; Perry

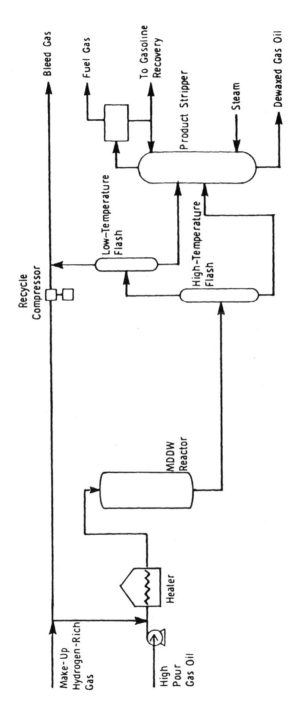

Figure 5.19 MDDW process schematic. (From Wise et al., 1986.)

Table 5.13 Pilot Plant MDDW
Yield, Arab Light Atmospheric Gas
Oil Charge

Distillate pour point: $-25°C$

	Yield, wt %
Ethane and lighter	0
Propane	1.3
Butanes	3.3
Pentanes	2.9
C_6-165°C gasoline	5.9
165°C$^+$ distillate	86.4
Gasoline, R + 0	90

Source: Donnelly and Green (1980).

et al., 1978; Ireland et al., 1979; Graven and Green, 1980; Donnelly and Green, 1980). A typical product yield from an Arab Light Atmospheric Gas Oil (24.1 API Gravity and +15°C pour point) is shown in Table 5.13. The byproduct gasoline, which represents two-thirds of the cracked product, as a high enough octane rating to be blended directly into the gasoline pool. Of course, distillate product yield decreases and gasoline yield increases as the wax content of the charge stock increases, and also as the pour point of the product decreases.

Feedstock to the MDDW process does not require hydrotreating because the catalyst can selectively exclude most of the nitrogen and sulfur compounds, some of which are known as catalyst poisons. After going on-stream the reactor temperature required to produce target pour product initially increases fairly rapidly to a line-out temperature.

Historically, cold flow properties determine the end point of distillate fuels, but over the years the maximum end point became an independent specification in many countries. Depending on the crude source and the climate conditions of the markets served, many refineries must undercut the end point of No. 2 fuel oil and diesel fuel below specification boiling point limits in order to meet fluidity requirements. With MDDW, not only would undercutting be unnecessary, high economic incentive exists for converting a portion of the heavy distillate from the residual fuel pool into marketable distillate fuels. Thus, this new technology offers an opportunity to untie the historical relationship between cold flow properties and end point of distillate fuels and to increase the amount of high-value distillates with acceptable pour points producible from a given crude, if the end point of the product is relaxed. Extensive field tests of extended boiling range, dewaxed distillates conducted in various European countries (Chen et al., 1977; Ireland et al., 1979) showed similar combustion performance to conventional diesel fuels and heating oils. So far, end point relaxation has taken place in European countries, but not in the United States.

By removing the wax from the high pour point distillate, its pour point blending characteristic is changed from that of virgin gas oils. It is well known that the pour point of virgin distillates blends nonlinearly (Reid and Allen, 1951), i.e., the pour point of a blend is higher than that of a calculated weighted linear average value. When a high pour virgin distillate is dewaxed and then blended with a low pour point virgin distillate, the blend has an expected lower pour point than that a calculated weighted linear average. This is illustrated in Figure 5.20 (Chen et al., 1978). Advantage can be taken of the favorable blending characteristic of catalytically dewaxed oils by fractionating a gas oil, dewaxing only the higher boiling fraction, and then blending these two streams.

In addition to processing high pour point virgin gas oils, the MDDW process may be integrated with catalytic cracking process especially for

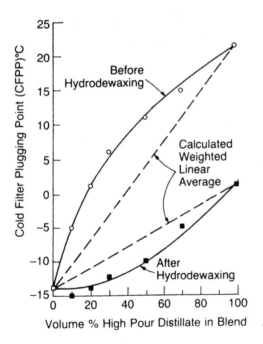

Figure 5.20. Synergistic pour point blending of hydrodewaxed high pour point distillate. (From Chen et al., 1978.)

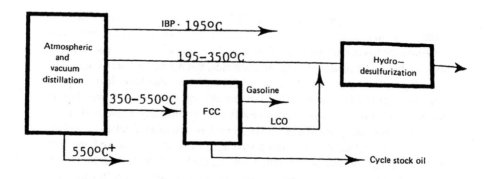

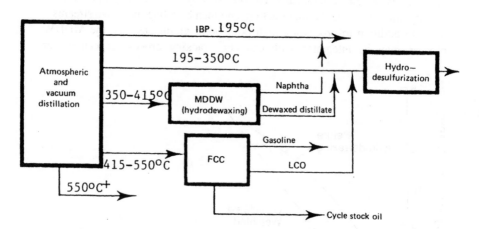

Figure 5.21   Distillate fuel manufacturing schemes. (From Perry et al., 1978.)

processing gas oils from very waxy crudes by either dewaxing of FCC cycle oils or diverting the light portion of the catalytic cracker feed to the MDDW, unloading the catalytic cracker for processing heavier feed (Perry et al., 1978; Pavlica et al., 1980; Donnelly and Green, 1980; Graven and Green, 1980). This is particularly desirable when the market for distillate fuel exceeds that of the gasoline market.

One such processing scheme is shown schematically in Figure 5.21. In this scheme, the 350–415°C fraction of the FCC feedstock is diverted to the MDDW unit. This allows the FCC unit to operate at higher recycle rates and higher distillate yields. The gasoline-to-distillate ratio of a typical FCC unit operating in the distillate mode may be as low as 1.3. With the addition of a

Table 5.14 Impact of MDDW on Distillate-to-Gasoline Ratio

Feed boiling range, °C:	FCC	FCC + MDDW
FCC	350–500	410–550
MDDW	—	350–410
Gasoline/distillate ratio	1.3	0.7
Distillate fuel composition, vol%		
light virgin gas oil	81	59
FCC light cycle oil	19	23
MDDW product	—	18

Source: Perry et al. (1979).

MDDW unit, laboratory data shows that it is possible to reduce the gasoline-to-distillate ratio produced to below 0.7 (Table 5.14).

Of course in order to realize the full benefit of this scheme, it is necessary to relax the end point of the distillate fuel, but not sacrificing the low temperature fluidity requirements.

When processing high-sulfur, high-nitrogen distillates (Gorring and Shipman, 1975), it is preferable to dewax the gas oil first and cascade the entire effluent from the dewaxing reactor to a conventional hydrotreating reactor to avoid the adverse effect of ammonia and hydrogen sulfide on the dewaxing catalyst and the need to have a two-stage process with interstage separation. The cascade process saves a significant amount of process energy (Shen, 1983) over the two-stage process.

The efficiency of the freezing point/pour point lowering over ZSM-5 increases with conversion (Chen et al., 1979). This nonlinear relationship between pour point and conversion is illustrated in Figure 5.22. Complete extraction of the 12% n-paraffins in the charge by shape selective sorption in 5 Å zeolite lowered the pour point to only $-18°C$. Since more than 12% of the charge is being converted over ZSM-5 to reduce its pour point to $-18°C$ and lower, other high pour point molecules, primarily isoparaffins and a lesser amount of selected 1-ring aromatics and naphthenes, are converted.

At $-25°C$ pour, only a trace of n-paraffins remained (Table 5.15). Once the n-paraffins are gone, the pour point dropped dramatically with only a small additional conversion to C_{17}^- product. This is the subtle change that accounts for the nonlinearity of pour point with conversion.

Selected alkylbenzenes and alkylnaphthenes are converted to lower boiling range and appear as a high octane naphtha product (Table 5.16).

Figure 5.23 shows a more detailed pattern of conversion of various classes of paraffins and alkylbenzenes as a function of the pour point of the

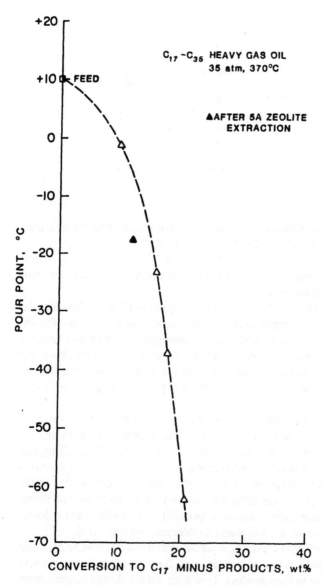

Figure 5.22 Pour point versus conversion. (From Chen et al., 1979).

Table 5.15 Sensitivity of Pour Point to Paraffin Content

C_{17}–C_{25} heavy gas oil, 35 atm, 370°C

	Charge				
Pour point, °C	+10	0	−25	−40	−60
Paraffin content, wt %	36.8	30.8	25.2	23.6	21.8
n-Paraffins detectable by M.S.	Yes	Yes	Trace	No	No
Incremental pour point lowering percent paraffin loss	—	3.3	7.1	15.6	25.0

Source: Chen et al. (1979).

Table 5.16 Compositional Change Before and After Dewaxing

Feed: C_{17}–C_{25} heavy gas oil

	Feed +10°C pour	Product, 69 wt % −60°C pour	wt % converted
Aromatics, wt %	26.9	36.7	5.9
alkylbenzenes	6.1	7.1	20
tetralins	5.8	8.2	2
naphthalenes	3.5	4.9	3
acenaphthenes	4.1	5.8	2
high polynuclears	7.4	10.7	Nil
Naphthenes	35.8	45.6	11.9
1-ring	17.4	20.4	19
2-ring	9.7	13.1	7
3-ring	4.6	6.2	6
4–5-ring	4.1	6.0	Nil
Paraffins	37.3	17.6	67.4
C_{17}	4.1	1.4	76
C_{18}	3.7	1.5	72
C_{19}	5.5	0.6	92
C_{20}	5.2	3.2	58
C_{21}	5.1	4.1	45
C_{22}	4.4	1.6	75
C_{23}	3.0	1.8	59
C_{24}	2.7	1.0	74
C_{25}	1.6	1.2	48
C_{26}^+	2.0	1.2	59

Source: Chen et al. (1979).

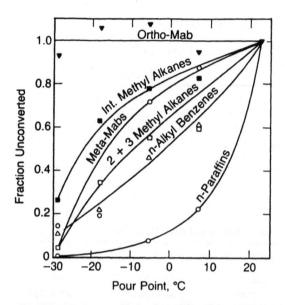

Figure 5.23 Conversion of alkane and alkylbenzene molecular classes during dewaxing of chargestock A over ZSM-5. Mab = methylalkylbenzene. (From Bendoraitis et al., 1986.)

dewaxed product. Examination of the trends for paraffin conversion shows that all *n*-paraffins in the gas oil (boiling range 310–455°C) are readily converted, followed by monomethyl paraffins. Paraffins containing quaternary carbons or polymethyl substituted are not converted. The conversion pattern of the methyl alkylbenzenes indicates a sharp cut-off between the meta and ortho isomers, which differ less than 0.5 Å in their critical molecular dimensions.

C. Jet Fuel Dewaxing

Jet fuels are a special type of distillate fuels high in flash point (>170°C) to reduce fire hazard during refueling, high in heating value for long flying range, and low in freezing point for high altitude. Traditionally, they were produced from petroleum by distillation and purification with little boiling range conversion or catalytic upgrading. Thus, by necessity, they are crude-dependent and are limited to very narrow boiling ranges. For example, the commercial jet fuel Jet A has a boiling range of 170 to 270°C and a freezing point of −40°C, and the premium military jet fuel JP-7 specifies an even narrower boiling range because it also requires a minimum heating value of 18,750 BTU/lb, which lim-

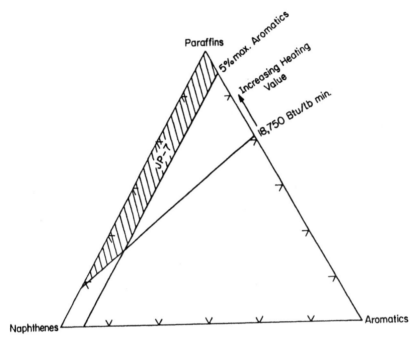

Figure 5.24   Chemical composition of JP-7 jet fuel (vol %). (From Chen and Garwood, 1987.)

its its aromatics content to 5 vol % maximum. To meet these requirements, its hydrocarbon composition is confined to a relatively narrow region rich in paraffins, as shown by Figure 5.24. By combining catalytic dewaxing with product hydrogenation, it is now possible to produce these jet fuels from a wide range of feedstocks.

Using the small pore zeolite Ni-erionite, described previously in Section III.B, jet fuel specification product has been made from a −40°C freezing point paraffinic feed. However, the dewaxing reaction was found to be dependent on hydrogen pressure (Chen and Garwood, 1978b). Freezing point lowering was effected at >69 atm of hydrogen pressure (Table 5.17). The same feed processing over ZSM-5 requires only a moderate pressure of 34 atm.

As shown in Figure 5.25, the freezing point of the jet fuel product from the ZSM-5 processing, as in the case of pour point lowering of distillate fuels, decreased nonlinearly with the extent of boiling range conversion. At low conversions, the freezing point of the product decreased only slightly. But as the

Table 5.17 Effect of Pressure on Freeze Point of Jet Fuel

Ni/H-erionite, 400°C, 14 LHSV, 27/1 H₂/hydrocarbon				
Pressure, atm	14.6	35	69	137
Conversion, wt %				
n-paraffins	15.6	16.2	58.2	73.8
non-normals	12.1	18.2	21.8	22.7
Jet fuel yield, wt %	87.0	82.3	69.1	64.6
Freeze point, °C	−38	−39	−47	−53
Δ Freeze point	+2	+1	−7	−13
Freeze point depression efficiency °C/% conversion			0.23	0.37

Source: Chen and Garwood (1987, 1978a).

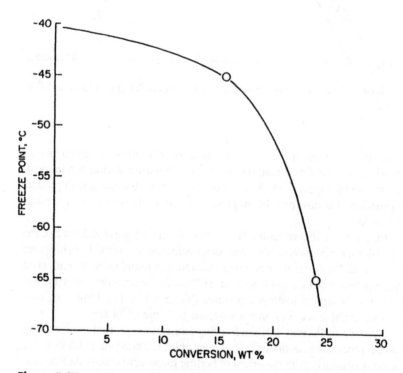

Figure 5.25 Catalytic cracking of jet fuel. (From Chen and Garwood, 1987.)

Table 5.18 Catalytic Dewaxing of 177–260°C and 260–290°C Kerosene Fractions

Zn/H-ZSM-5, 34 atm, 24 LHSV

	Charge stock:	
	177–260°C (Stock B)	260–290°C (Stock C)
H₂/HC mole ratio	5	7
Temperature, °C	340	370
Conversion, wt %	13.4	22.0
Jet fuel product		
freezing point, °C	−65	−43
KV at −35°C, cs	8.52	20.70

Source: Chen and Garwood (1987).

freezing point was lowered to below −40°C, each additional percent increase in conversion led to more than 2.3°C drop in freezing point.

With catalytic dewaxing, it is also possible to convert a portion of the heavier distillate to jet fuels (Chen and Garwood, 1986). For example, the supply of Jet A could be increased 25% or more by the inclusion of 260–290°C boiling range kerosene. Shown in Table 5.18 are the results of dewaxing a 177–260°C and a 260–290°C kerosene fractions. By blending the high-boiling product with the standard fuel, potentially all of the 260–290°C fraction available from most crudes could be upgraded to jet fuel and meet all the quality requirements.

Premium military jet fuel, JP-7, requires a ≤ −45°C freezing point and a minimum heating value of 18,750 BTU/lb, which can be met only with paraffinic stocks. Waxy crudes such as the Libyan crude, being highly paraffinic, are potential sources for the manufacturer of high-energy jet fuels except for their high freezing point. With catalytic dewaxing over ZSM-5 combined with product hydrogenation, JP-7 specification fuel can be produced. A comparison of JP-7 specification with the product from the combined catalyst system is shown in Table 5.19.

D. Mobil Lube Dewaxing Process

Mobil's Lube Dewaxing (MLDW) process differs from the fuels dewaxing process in that it utilizes a two-reactor system. A schematic process flow diagram is shown in Figure 5.26. The first reactor contains the ZSM-5–based

Table 5.19 Catalytic Dewaxing Followed by Hydrogenation

	Libyan 177–260°C kerosene			
	Charge stock E	Dewaxed product[a]	Then hydrogenated[b]	JP-7 specification
Gravity, °API	48.1	47.2	47.5	44–50
Freezing point, °C	−33	−53	−54	−45 max
Aromatics, vol %	9.1	11.9	4.0	<5
Heating value, BTU/lb	18,710	18,655	18,835	18,750 min

[a]Zn/H-ZSM-5, 343°C, 34 atm, 15 H_2/HC mole ratio, 24 LHSV. 78.5 wt % yield of dewaxed product.
[b]Ni/Kieselguhr, 34 atm, 315°C.
Source: Chen and Garwood (1987).

catalyst. The effluent from the first reactor is cascaded to the second reactor without product separation. The second reactor contains a hydrofinishing catalyst to assure that the final lube product will meet all quality and engine performance specifications (Graven and Green, 1980; Smith et al., 1980).

For solvent-refined raffinates the catalyst may age at one or more degrees centigrade per day, making the process cyclic with cycle lengths lasting several weeks to several months. For hydrotreated feeds, the process is operated in the steady-state mode with cycle length exceeding a year.

Chen et al. (1991) studied the mechanism of catalyst deactivation in MLDW with varying feedstocks and at varying LHSV. Nitrogen content and the average molecular weight of the feed were found to be related to deactivation rate. The following deactivation model was found to be applicable to the experimental data:

$$\frac{\text{Deactivation rate}_1}{\text{Deactivation rate}_2} = \left[\frac{\text{LHSV}_2}{\text{LHSV}_1}\right]^{E_d/E_a}$$

where E_d/E_a is the ratio of activation energy of deactivation to the activation energy of the dewaxing reaction. This ratio turned out to be a temperature-sensitive function. If it is greater than 1, then the rate of deactivation is more sensitive than the rate of dewaxing when the reaction temperature is increased; the opposite is true when the ratio is less than 1. For light neutrals, this ratio was generally less than 1. The model predicts that the catalyst should deactivate gradually toward a steady-state activity, which was consistent with commercial data.

The process can be used to produce a full slate of lube basestocks, ranging from the conventional lubes and hydrocracked lubes to very low pour point

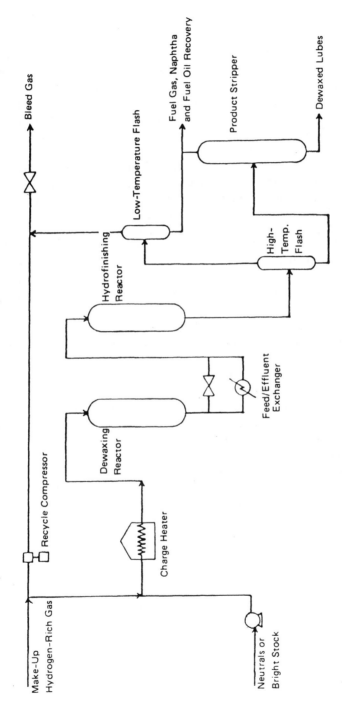

Figure 5.26 Simplified MLDW process flow diagram. (From Graven and Green, 1980.)

products required for transformer and refrigeration services. It offers many advantages over the solvent dewaxing technology, including lower capital investment and lower operating cost. The feed to MLDW can be properly stripped raffinates (Gillespie et al., 1979, 1980, 1984) from any of the solvent extraction processes (furfural, phenol, *N*-methyl-2-pyrrolidone) commonly practiced in lube manufacture to remove aromatics, or raw distillates, soft waxes, and deasphalted resids.

The combination of commercial lube hydrotreating process and MLDW offers for the first time the potential of developing an all-catalytic route to lubes. Several companies that have such hydrotreating facilities have added catalytic dewaxing for their lighter stocks (Garwood and Silk, 1981; Wise et al., 1986).

If economic incentives are great enough to justify wax production, a combination of solvent dewaxing with MLDW can be attractive (Chen and Garwood, 1973b; Smith et al., 1980). Saleable wax can be recovered by partial solvent dewaxing at an intermediate temperature above the freezing point of water. Partial solvent dewaxing is much less costly than complete solvent dewaxing because of higher filter rates, easier wax deoiling, and reduced refrigeration requirements. Foots oil, a waxy byproduct of solvent dewaxing, can also be suitable stock for catalytic dewaxing (Garwood and Wise, 1976). MLDW is suitable for a wide variety of stocks, from solvent extracted raffinates to hydrocracked distillates, from spindle oils to brightstocks, and in some instances unrefined gas oils (Baker and McGuiness, 1995).

Typical lube yields from the MLDW process are shown in Table 5.20. The yields are based on processing an Arab Light crude. The byproduct yield is similar to that of the distillate dewaxing process, i.e., one-third LPG and two-thirds naphtha.

Physical properties of catalytically dewaxed basestocks are similar to those of the solvent dewaxed stocks with the exception of viscosity and viscosity index (VI). VI is traditionally the parameter used to control the severity

Table 5.20 Commercial MLDW Lube Yields

Feed: Arab Light raffinates	
	Yield, wt %
100 SUS light neutral	74.5
300 SUS heavy neutral	79.8
700 SUS heavy neutral	82.2
Bright stock	91.2

Table 5.21 Engine Oil Quality Tests

| | Marine diesel engine oil: | | |
	SDW	MLDW	
Caterpillar 1-G			
total weighted demerits	70	47	100 max
total groove fill, %	18	18	25 max
Passenger car engine oil			
(Ford MS sequence VC):			
sludge	8.5	9.5	8.5 min
varnish	8.2	8.8	8.0 min
piston skirt varnish	8.0	8.3	7.9 min
cold cranking simulator cp at −18°C	2400	2400	2400 max

Source: Smith et al. (1980).

of the aromatics extraction step that precedes dewaxing. With the same extraction severity, the viscosity of the MLDW product is higher and the viscosity index lower than that of the SDW product. These differences do not, however, adversely, affect end-use performance of the finished oils. The difference decreases with increasing basestock viscosity and is negligible with bright stocks.

Extensive product quality tests have been conducted on various finished products, including engine oils, industrial turbine and circulating oils, gear oil, and refrigerator and transformer oils, made by blending catalytic dewaxed basestocks, comparing them with those made with solvent dewaxed basestocks. In all tests, products containing the catalytic dewaxed oil blends show equal or better performance than their solvent dewaxed counterparts. Table 5.21 shows the test results for engine oils.

Sarli and Bortz (1991) studied the production of low pour point transformer oils from Arabian Light distillates using the MLDW process and obtained better than 65% yield of −46°C pour point product.

The MLDW process was tested on a large scale in Mobil's Gravenchon Refinery in France in 1978. The first grass-root commercial MLDW unit went on stream in 1981. A 15,000 B/D unit at Mobil's Paulsboro Refinery processes a full range of lube basestocks. There are now nine licensed units operating worldwide with an installed capacity of 50,650 B/D, including in the United States, Australia, and Japan.

E. Development of Bifunctional Catalytic Dewaxing Process

Other zeolites such as the synthetic ferrierite, a dual pore zeolite with interconnecting channels of elliptical 10- and 8-membered oxygen ring openings, and mordenite, a dual pore zeolite with interconnecting channels of puckered 12- and 8-membered oxygen ring openings, have also been found to be effective lube dewaxing catalysts with the incorporation of a noble metal hydrogenative function (Winquist, 1982; Bennett et al., 1975). Because their larger channels are undirectional and the smaller channels are accessible only by the smaller molecules, they are not as effective as the medium pore zeolites for heavier feedstocks.

The catalytic dewaxing process developed by the British Petroleum Company (Donaldson and Pout, 1972; Bennett et al., 1975; Hargrove et al., 1978) is a catalytic hydrocracking process which employs a bifunctional platinum/H-mordenite catalyst (Burbidge et al., 1971). The fixed-bed process is operated under high hydrogen pressure to saturate the cracked byproducts consisting largely of propane, butanes, and pentanes. High hydrogen pressure also minimizes coke formation and maintains stable catalyst activity. This is particularly important with mordenite, which contains unidirectional channels prone to coking when used as an acidic catalyst. The catalyst has very little activity for desulfurization or denitrogenation.

The process is applicable to the production of low pour point distillate fuels and very low pour point low viscosity oils from naphthenic and partially dewaxed paraffinic distillates, but is not suitable for the production of high viscosity index, high viscosity lubricating oils. Dewaxed products with pour point as low as $-50°C$, suitable as transformer and refrigerator oils, and low pour point extended boiling range diesel fuel and No. 2 heating oils have been produced.

A similar catalytic dewaxing process employing a bifunctional Pd/H-ferrierite catalyst was developed by Shell (Winquist, 1982). The process can be combined with a solvent dewaxing process using methyl isopropyl ketone as the solvent, which produces an intermediate pour point stock for the catalytic dewaxing process (Stem, 1986).

As discussed in Chapter 4, Section II.B, more recent studies on acid/metal balanced large pore and medium pore zeolites have shown their hydroisomerization selectivity to be significantly improved over the monofunctional catalysts. The hydroisomerization activity of a catalyst consisting of a strong hydrogenation metal such as platinum on an acidic support is well known, and hydroisomerization is a part of the dual functional mechanism of hydrocracking (Coonradt and Garwood, 1964). Based on these studies, several bifunctional catalytic dewaxing processes have been developed.

Table 5.22 Comparison of MSDW and MLDW in Dewaxing a hydrocracked Persian Gulf Gas Oil

	MSDW	MLDW
Lube pour point, °C	−15	−15
Lube yield, wt %	85.7	81.7
Lube kV at 40°C, cSt	30.5	37.0
Lube VI	102	97

Source: Baker and McGuiness (1995).

Mobil has developed several lube dewaxing processes involving hydroisomerization, including the Mobil Selective Dewaxing (MSDW) process, Mobil Wax Isomerization (MWI) process (Baker and McGuiness, 1995), and the Mobil Isomerization Dewaxing (MIDW) process (Buyan et al., 1995).

Mobil's MSDW process uses a more selective catalyst than the ZSM-5 catalyst, designed for a class of "clean" feedstocks, i.e., low in sulfur, nitrogen and coke precursors. By incorporating a noble metal into the highly shape selective dewaxing catalyst increases catalyst activity and cycle length and the combination of hydroisomerization and selective cracking results in improved lube yield and viscosity index (VI). Results shown in Table 5.22 compare MSDW with MLDW at 28 atm of a 150N distillate derived from hydrocracking a Persian Gulf vacuum gas oil.

Mobil's MWI process converts high wax content streams into very high VI lubes. The process consists of a two reactor mild hydrocracking/hydroisomerization system followed by distillation to remove light ends. A solvent dewaxer is used to reduce the pour point to specification values.

Mobil's MIDW process operates at mild conditions similar to those used in light gas oil desulfurization units. It can operate at low hydrogen pressures, not more than 35 atm, 260–440°C and 1 LHSV. The process has been in commercial operation. A typical example of the MIDW commercial product properties is shown in Table 5.23. The naphtha, low in sulfur, is ready for the catalytic reformer without pretreatment. The kerosene is a high quality jet fuel blending component, as indicated by the high smoke point of 28 mm, low sulfur content of 2 ppmw, low freeze point of less than −54°C and the low aromatics content of less than 10%. The diesel fuel is also very high in quality. The low sulfur LSFO is very low in pour point and slightly enriched in naphthenes and aromatics. The process may also be applied to process slack wax to produce high quality lubricating oils (Buyan, F. M. et al., 1995).

Chevron has developed its isodewaxing process based on a bifunctional catalyst (Miller, 1989, 1993). They combine their Isodewaxing with Chevron's

Table 5.23 MIDW Commercial Product Properties

	Feedstock	Naphtha	Kerosene	Diesel	LSFO[a]
TBP cut, °C	340–510	C$_5$-150	150–255	255–390	390+
API gravity	32.0	73.0	49.5	34.7	29.5
Specific gravity at 15°C	0.815	0.695	0.782	0.851	0.879
Sulfur, ppmw	260	<1	2	<20	40
Smoke point, mm	—	—	28	—	—
Freeze point, °C	—	—	<−54	—	—
Pour point, °C	>40	—	—	−43	<−54
Cetane index	—	—	52	56	—
P/N/A,[b] wt %	44/39/17	44/45/22	—/—/<10	45/31/24	36/42/22

[a]LSFO, low sulfur fuel oil.
[b]P/N/A, paraffin/naphthene/aromatics.
Source: Buyan, F. M., N. Y. Chen, R. B. LaPierre, D. A. Pappal, R. D. Partridge, and S. F. Wong, Hydrocarbon Tech. Intern. *Autumn*, 14 (1995).

Hydrogen Pretreating and Hydrofinishing for lube oil processing. The process operates at 150 atm, 365–400°C, and 1 LHSV.

Hydrogen pretreating is used to lower nitrogen and sulfur content in the feedstock enabling the isomerization and hydrofinishing catalysts to last longer and maximize the yield. Their dewaxing catalyst isomerizes the waxy feedstock to produce a high-yield, high-VI, and low–pour point basestock, and the byproducts include high quality aviation fuel and diesel fuel. The hydrofinishing step also uses Chevron's Hydrofinishing catalyst to produce extremely stable, white as water basestocks.

REFERENCES

Anderson, C. D., F. G. Dwyer, G. Koch, and P. Niiranen, "A New Cracking Catalyst for Higher Octanes Using ZSM-5," paper presented at the 9th Iberoamerican Symp. Catal., Lisbon, Portugal, July, 1984.

"ASTM Manual for Rating Motor, Diesel and Aviation Fuels," ASTM, Philadelphia, (1973).

Baker, C. L. and M. P. McGuiness, "Mobile Lube Dewaxing Technologies," paper presented at the 1995 NPRA. Annual Meeting, San Francisco (AM-95-56), March 19–22, 1995.

Bell, A. T., L. E. Manzer, N. Y. Chen, V. W. Weekman, L. L. Hegedus, and C. J. Pereira, Chem. Eng. Prog. *91*(2), 26 (1995).

Bendoraitis, J. G., A. W. Chester, F. G. Dwyer, and W. E. Garwood, "Pore size and Shape Effects in Zeolite Catalysis," paper presented at the 7th International Zeolite Conference, Tokyo, August 17–22, 1986.

Bennett, R. N., G. J. Elkes, and G. J. Wanless, Oil Gas J. *3(1)*, 69 (1975).

Benson, J., Chemtech *6*, 16 (1976).

Bernard, J. R., J. Bousquet, and M. Grand, German Pat. 2,626,840, Dec. 23, 1976.

Bonacci, J. C. and J. R. Patterson, U.S. Pat. 4,292,167, Sept. 29, 1981.

Buchanan, J. S., Appl. Catal. *74*, 83 (1991).

Burbidge, B. W., I. M. Keen, and M. K. Eyles, Advan. Chem. Ser. *102*, 400 (1971).

Burd, S. D. and J. Maziuk, Oil Gas J. *70(27)*, 52 (1972).

Buyan, F. M., N. Y. Chen, R. B. LaPierre, D. A. Pappal, R. D. Partridge, and S. F. Wong, Hydrocarbon Techn. Intern. Autumn Quarterly 1995, p. 14.

Chen, N. Y, U.S. Pat. 3,729,409, Apr. 24, 1973.

Chen, N. Y and W. E. Garwood, U.S. Pat. 3,379,640, Apr. 23, 1968.

Chen, N. Y and W. E. Garwood, Advan. Chem. Ser. *121*, 575 (1973a).

Chen, N. Y and W. E. Garwood, U.S. Pat. 3,755,138, Aug. 28, 1973b.

Chen, N. Y and W. E. Garwood, Ind. Eng. Chem. Process Des. Dev. *17*, 513 (1978b).

Chen, N. Y and W. E. Garwood, J. Catal. *52*, 453 (1978a).

Chen, N. Y and W. E. Garwood, Ind. Eng. Chem. Prod. Res. Devel. *25*, 641 (1986).

Chen, N. Y and E. J. Rosinski, U.S. Pat. 3,630,966, Dec. 28, 1971.

Chen, N. Y, J. Maziuk, A. B. Schwartz, and P. B. Weisz, Oil Gas J. *66(47)*, 154 (1968).

Chen, N. Y, R. L. Gorring, H. R. Ireland, and T. R. Stein, Oil Gas J. *75(23)*, 165 (1977).

Chen, N. Y, B. M. Gillespie, H. R. Ireland, and T. R. Stein, U.S. Pat. 4,067,797, Jan. 10, 1978.

Chen, N. Y, W. E. Garwood, W. O. Haag, and A. B. Schwartz, "Shape Selective Hydrocarbon Catalysis over Synthetic Zeolite ZSM-5," paper presented at Symp. Advan. Catal. Chem. I, Snowbird, UT, Oct. 3–5, 1979.

Chen, N. Y, W. E. Garwood, and R. H. Heck, Ind. Eng. Chem. Process Des. Dev. *26*, 706 (1987).

Chen, N. Y, A. Y. Kam, C. R. Kennedy, A. B. Ketkar, D. M. Nace, and R. A. Ware, U.S. Pat. 4,740,292, Apr. 26, 1988.

Chen, N. Y, W. E. Garwood, and S. B. McCullen, U.S. Pat. 4,814,543, Mar. 21, 1989.

Chen, N. Y., T. F. Degnan, J. D. Lutner, and B. P. Pelrine, Stud. Surf. Sci. Catal. *68*, 773 (1991).

Coonradt, H. L. and W. E. Garwood, I&EC Proc. Des. Dev. *3*, 38 (1968).

Donaldson, K. and C. R. Pout, "The Application of a Catalytic Dewaxing Process to the Production of Lubricating Oil Basestocks," paper presented at the 164th American Chemical Society National Meeting, New York, Sept. 1972.

Donnelly, S. P. and J. R. Green, "Recent MDDW Process Improvements," paper presented at the Symp. Jap. Petrol. Inst., Tokyo, Oct. 2–3, 1980.

Donnelly, S. P., S. Mizrahi, P. T. Sparrell, A. Huss, P. H. Schipper, and J. A. Herbst, "How ZSM-5 Works in FCC," paper presented at the American Chemical

Society, New Orleans Aug. 30–Sept. 4, 1987, Div. of Pet. Chem. Preprints, pp. 621–626.

Dwyer, F. G., P. H. Schipper, and F. Gorra, "Octane Enhancement in FCC via ZSM-5," paper presented at the 1987 NPRA Annual Meeting, San Antonio, TX (AM-87-63), March 30, 1987.

Evitt, S. D., F. Gong, M. N. Harandi, and H. Owen, "New Zeolite Based Technologies for Benzene and Olefin Reduction in Gasoline," paper presented at the 1992 NPRA Annual Meeting, New Orleans (AM-92-55), March 22–24, 1992.

Fromager, N. H., J. R. Bernard, and M. Grand, Chem. Eng. Prog. *73(11)*, 89 (1977).

Garwood, W. E., Am. Chem. Soc. Symp. Ser. *218*, 383 (1983).

Garwood, W. E. and N. Y. Chen, "Octane Boosting Potential of Catalytic Processing of Reformate over Shape Selective Zeolite," paper presented at the American Chemical Society Meeting, Houston, TX, Mar. 23–28, 1980; Div. Petrol. Chem. Preprint *25(1)*, 84 (1980).

Garwood, W. E. and M. R. Silk, U.S. Pat. 4,283,271, Aug. 11, 1981; U.S. Pat. 4,283,272, Aug. 11, 1981.

Garwood, W. E. and J. J. Wise, U.S. Pat. 3,960,705, Jun. 1, 1976.

Goelzer, A. R., A. Ram, A. Hernandez, A. A. Chin, M. N. Harandi, and C. M. Morris, "Mobil-Badger Technologies for Benzene Reduction in Gasoline," paper presented at the 1993 NPRA Annual Meeting, San Antonio, TX (AM-93-19), March 21–23, 1993.

Gorring, R. L. and G. F. Shipman, U.S. Pat, 3,894,938, Jul. 15, 1975.

Gillespie, B. M., M. S. Sarli, and K. W. Smith, U.S. Pat. 4,137,148, Jan. 30, 1979; U.S. Pat. 4,181,598, Jan. 1, 1980; U.S. Pat. 4,437,975, Mar. 20, 1984.

Graven, R. G. and J. R. Green, "Hydrodewaxing of Fuels and Lubricants Using ZSM-5 Type Catalysts," paper presented at the 1980 Congr. Australia Inst. Petrol., Sydney, Sept. 15–17, 1980.

Harandi, M. N., D. L. Johnson, J. D. Kushnerick, H. Owen, P. H. Schipper, and S. Yurchak, "Light Olefin Upgrading to Gasoline Using the Mobil Olefins-to-Gasoline (MOG) Process," paper presented at the AIChE 1991 Spring National Meeting, Houston, TX, Apr. 7–11, 1991.

Hargrove, J. D., G. J. Elkes, and A. H. Richardson, "BP Cat. Dewaxing—Experience in Commercial Operation," paper presented at the NPRA National Fuels and Lubricants Meeting, Houston, TX, Nov. 9–10, 1978.

Heck, R. H. and N. Y. Chen, Ind. Eng. Chem. Res. *32*, 1003 (1993).

Heinemann, H., Catal. Rev.-Sci. Eng. *15*, 53 (1977).

Ireland, H. R., C. Redini, A. S. Raff, and L. Fava, Hydrocrabon Process. *58(5)*, 119 (1979).

Johnson, T. E. and A. A. Avidan, "FCC Design for Maximum Light Olefin Production," paper presented at the 1993 NPRA Annual Meeting, San Antonio, TX (AM-93-51), Mar. 21–23, 1993.

Kunchal, S. K., F. G. Dwyer, and C. P. Ambler, "Isofin—A New Process for Olefin Isomerization," paper presented at the 1993 NPRA Annual Meeting, San Antonio, TX (AM-93-45), Mar. 21–23, 1993.

Liers, J., J. Meusinger, A. Mösch, and W. Reschetilowski, Hydrocarbon Process. *72*(8), 165 (1993).

Madon, R. J., J. Catal., *129*, 275 (1991).

Meisel, S. L., J. P. McCullough, C. H. Lechthaler, and P. B. Weisz, "Recent Advances in the Production of Fuels and Chemicals Over Zeolite Catalysts," paper presented at the Leo Friend Symposium, 174th Am. Chem. Soc. Mtg., Chicago, Aug. 30, 1977.

Miller, S. J., U.S. Pat. 4,859,311, Aug. 22, 1989.

Miller, S. J., "New Molecular Sieve Process for Lube Dewaxing by Wax Isomerization," paper presented at the Symposium on New Catalytic Materials utilizing Molecular Sieves, Div. of Petrol. Chem., Am. Chem. Soc. Chicago Mtg., Aug. 22–27, 1993.

Pavlica, R. T., T. R. Stein, and C. W. Streed, U.S. Pat. 4,192,734, Mar. 11, 1980.

Perry, R. H., Jr., F. E. Davis, Jr., and R. B. Smith, Oil Gas J. *76*(*21*), 78 (1978).

Plank, C. J., E. J. Rosinski, and E. N. Givens, U.S. Pat. 4,141,859, Feb. 27, 1979.

Plank, C. J., E. J. Rosinski, and E. N. Givens, U.S. Pat. 4,276,151, Jun. 30, 1981.

Ram, S. and A. A. Chin, "Mobil's New MOI Process for Olefin Interconversion," paper presented at the Worldwide Solid Acid Process Conf. Houston, TX, Nov. 16, 1993.

Ramage, M. P., K. R. Graziani, P. H. Schipper, and F. J. Krambeck Advan. Chem. Eng. *13*, 193 (1987).

Reid, E. B. and H. I. Allen, Petrol. Refiner *30*(*5*), 93 (1951).

Roselius, R. R., K. R. Gibson, R. M. Ormiston, J. Maziuk, and F. A. Smith, "Rheniforming and SSC, New Concepts and Capabilities," paper presented at the NPRA Annual Meeting, San Antonio, TX, Apr. 1–3, 1973.

Rosinski, E. J. and A. B. Schwartz, U.S. Pat. 4,309,280, Jan. 5, 1982.

Sarli, M. S. and R. W. Bortz, "Manufacture of Transformer Oil via the Mobil Lube Dewaxing Process," paper presented at the 1991 NPRA National Fuels and Lubricants Meeting, Houston, TX, Nov. 7–8, 1991.

Schipper, P. H., F. G. Dwyer, P. T. Sparrell, S. Mizrahi, and J. A. Herbst, "Zeolite ZSM-5 in Fluid Catalytic Cracking: Performance, Benefits, and Applications," Chapter 5, ACS Symp. Ser. *375*, 64, (1988).

Shen, R. C., U.S. Pat. 4,400,265, Aug. 23, 1983.

Smith, F. A. and R. W. Bortz, Oil Gas J. *88*(*33*) 51 (1990).

Smith, K. W., W. C. Starr, and N. Y. Chen, Oil Gas J. *78*(*21*), 75 (1980).

Stem, S. C., U.S. Pat. 4,622,130, Nov. 11, 1986.

Tabak, S. A., "Production of Synthetic Diesel Fuel from Light Olefins," paper presented at the AIChE National Meeting, Philadelphia, Aug. 1984.

Tabak, S. A., F. J. Krambeck, and W. E. Garwood, "Conversion of Propylene and Butylene over ZSM-5 Catalyst," paper presented at the AIChE Meeting, San Francisco, Nov. 25–30, 1984; AIChE J. *32*, 1526 (1986a).

Tabak, S. A., A. A. Avidan, and F. J. Krambeck, "Production of Synthetic Gasoline and Diesel Fuels from Non-petroleum Sources," paper presented at the ACS National Meeting, New York, Apr. 13–15, 1986b.

van den Berg, J., "Shell Details Olefin to 'Clean' Fuel Process," Europ. Comm. News, Jan. 21, 1991, p. 23.

van den Berg, J. P., H. W. Röbschläger, and I. E. Maxwell, "The Shell Poly-Gasoline and Kero (SPGK) Process," paper presented at the 11th North American Catalysis Society Zeolite Symposium, Dearborn, MI, May 7–11, 1989.

Weekman, V. W. and N. Y. Chen, "Catalysis for the Production of Fuels," paper presented at the North Am. Catal. Soc. Mtg., Lexington, KY, May 4–10, 1991.

Weisz, P. B. and J. R. Katzer, "Fuels from Carbonaceous Resources—Constraints and Opportunities," paper presented at CHEMRAWN III Conference, The Hague, Jun. 25–29, 1984.

Winquist, B. H. C., U.S. Pat. 4,343,692, Aug. 19. 1982.

Wise, J. J., J. R. Katzer, and N. Y. Chen, "Catalytic Dewaxing in Petroleum Processing," paper presented at the Am. Chem. Soc. 173rd Annual Mtg., New York, Apr. 14–15, 1986.

Yanik, S. J., E. J. Demmel, A. P. Humphries, and R. J. Campagna, Oil Gas J. *83*(*19*), 108 (1985).

Yurchak, S., J. E. Child, and J. H. Beech, "Mobil Olefins Conversion to Transportation Fuels," paper presented at the 1990 NPRA. Annual Meeting, San Antonio, TX (AM-90-37), March 25–27, 1990.

6

Applications in Aromatics Processing

I. BTX SYNTHESIS

A. M2-Forming

As discussed in Chapter 4, ZSM-5 converts light paraffins, olefins, and naphthenes to aromatics and light gases. A generic name, M2-Forming, has been coined to describe this new aromatization process (Chen and Yan, 1986). As shown by the data in Table 6.1, at 538–575°C and 1 WHSV, the aromatics yield from n-pentane and n-hexane is more than 30% by weight; similar yield is obtained from the more reactive propene at 68 LHSV. A virgin naphtha which contains 47% paraffins, 41% naphthenes, and 12% aromatics gives 44% aromatics, or a net gain of 32%. And a light FCC gasoline which contains 41% of olefins and 10% aromatics gives 54% aromatics, or a net gain of more than 44% aromatics.

In fact, all nonaromatic molecules are aromatizable. Only methane and ethane are not aromatized and their concentrations are continuously increased as the reaction severity is increased.

For paraffinic feed, the olefinic intermediates are mainly derived from the catalytic cracking step. With feedstocks containing olefins, diolefins, naphthenes, and hydroaromatic compounds, and at relatively low temperatures, direct aromatization via hydrogen transfer reactions may provide an alternate

Table 6.1 Aromatization of Light Hydrocarbons

| | | Catalyst: ZSM-5; pressure: atmospheric | | | |
| | | Reaction | | Atmospheric yield, % | |
Feed	C%	Temperature, °C	WHSV	observed	extinction
n-Pentane	83.33	575	1	31	45
n-Hexane	83.72	538	1	32	45
Propene	85.71	538	68	33	58

Source: Chen and Yan (1986).

route to aromatics and saturates. At higher temperatures, however, catalytic cracking again dominates the first reaction step.

M2-Forming uniquely produces aromatic concentrates. Refinery streams rich in unsaturated hydrocarbons, such as the byproduct streams from the naphtha steam cracking process known as pyrolysis gasoline or Dripolene, unsaturated from the catalytic cracking process and LPG are potential feedstocks for this process. Because of the high concentration of aromatics produced, the product stream may be separated directly for BTX production without a costly additional solvent extraction step. Thus, the M2-Forming process could be an attractive alternative to the conventional hydrotreating/extraction process for upgrading pyrolysis gasoline to BTX (Chen and Yan, 1986).

Depending on the reactivity of the feedstock, the operating temperature varies significantly, ranging from less than 370°C for olefins and 538°C for propane.

Compared with other zeolites, the HZSM-5 catalyst is unique in its ability to catalyze the aromatization reaction with sustained activity over a period of several days. Its stability is attributed to the shape and size of the pore openings and the tortuosity of the channels, which inhibit the formation of coke precursors (Chen and Garwood, 1978; Walsh and Rollmann, 1979). Nevertheless, because of eventual catalyst deactivation, periodic regenerations are necessary, thus the application of M2-Forming to aromatics production would be in the form of a cyclic process.

The introduction of gallium in zeolites has received considerable interest for the conversion of light paraffins to aromatics (Kitagawa et al., 1986; Price and Kanazirev, 1990; Abdul Hamid et al., 1994; Kwak et al., 1994). One such cyclic aromatization process, the Cyclar™ process, was announced by UOP and British Petroleum in 1984. A recent report (Oil Gas Journal, 1987) indicated that a first commercial-scale unit using the Cyclar process to convert LPG to aromatics, used as high octane gasoline blending components, will be built at the British Petroleum Grangemouth Refinery in Scotland. The process employs UOP's CCR reforming technology (continuous catalyst regeneration) to convert LPG to aromatics (Johnson and Hilder, 1984). Similar to naphtha reforming, the reaction is carried out in vertically stacked multistage adiabatic radial flow reactors with interstage heating. Figure 6.1 shows a schematic process flow diagram. In this process, the catalyst flows by gravity from one reactor to the next. After exiting the last reactor, the spent catalyst is transferred by lift gas to the top of the regenerator, where it is separated from the lift gas and flows by gravity again through the regenerator to a hopper. Lift gas is used to transfer the catalyst back to the top of the first reactor. The process was successfully demonstrated in 1990 but the unit was shut down in 1991 because of poor process economics (Euro. Chem. News, 1992).

Kitagawa et al. (1986) reported a yield of 71% aromatics from propane through the addition of gallium to the ZSM-5 catalyst. Others have also reported improved yields of aromatics from propane by adding a metal function to ZSM-5 (Davies and Kolombos, 1979; Anderson et al., 1985; Mole and Anderson, 1985). Some of the phosphate-based molecular sieves also have been

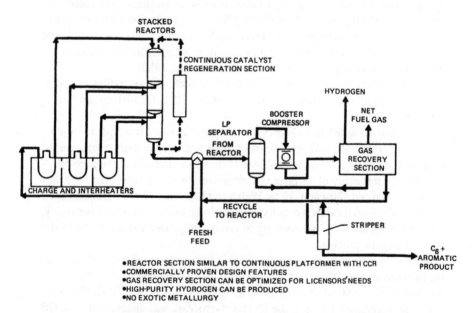

Figure 6.1 UOP/BP Cyclar process for LPG aromatization. (From Anderson et al., 1985.)

Table 6.2 Synthesis of Benzene from Light Naphtha

Catalyst			
Na, %	2.03	0.005	1.14
Pt, %	0.44	0.34	0.36
Re, %	0.42	0.37	0
Cl, %	0.49	<0.005	0
Products			
C_5^+, % of feed	52.67	37.54	48.55
Aromatics, % of C_5^+	88.63	92.05	92.82
Benzene, % of aromatics	89.14	34.12	94.76

Source: Detz and Field (1982).

reported to have aromatization activity for converting olefins and diolefins to BTX aromatics (Garska and Lok, 1985).

Detz and Field (1982) reported the selective synthesis of benzene from light naphtha (50–125°C) using a platinum-loaded low acidity ZSM-5. Table 6.2 shows the importance of lowering the sodium content and the chloride content to increase benzene selectivity.

B. Xylene Isomerization

The Mobil Vapor Phase Xylene Isomerization (MVPI) process is designed to isomerize the xylenes in a C_8 aromatics stream obtained from either reformates or pyrolysis gasolines. A schematic diagram of a xylene isomerization process is shown in Figure 6.2. The C_8 aromatics stream is fractionated in an ethylbenzene tower to remove ethylbenzene from the feed. Because the ethylbenzene separation step is energy intensive, in recent years it has often been omitted. Thus, the C_8 aromatics mixture charged to the isomerization reactor usually contains 10 to 40% ethylbenzene. After isomerization, in most cases, the *p*-xylene in the equilibrated mixture is separated by crystallization or molecular sieve adsorption, and the remaining liquid is recycled together with fresh feed to the isomerization reactor. In order to avoid the buildup of ethylbenzene concentration in the recycle loop, the ability of a catalyst to convert ethylbenzene with a minimal loss of xylenes is of critical importance to a satisfactory process.

The MVPI process operates by a different reaction pathway from that of the older processes such as octafining. The activity and stability of the MVPI catalyst and its product selectivity are the major advantages over the older processes. Octafining uses a noble metal containing bifunctional catalyst to

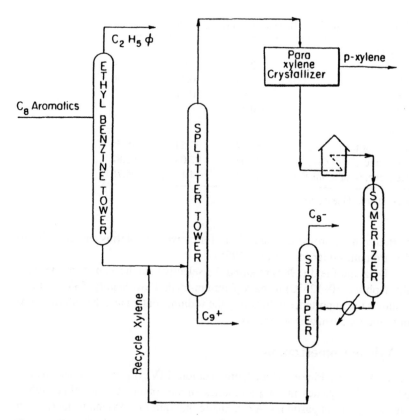

Figure 6.2 Schematic diagram of a xylene isomerization process. (From Haag and Olson, 1974.)

isomerize xylenes and convert ethylbenzene to xylenes (Figure 6.3). However, xylene losses are appreciable through other side reactions such as disproportionation and hydrogenolysis. The MVPI process uses a monofunctional acid catalyst. The process is carried out at 315–370°C. Ethylbenzene is converted to benzene and C_{10} aromatics by transalkylation reactions.

As mentioned, xylene isomerization over acid catalysts is also accompanied by side reactions such as disproportionation. Relative to larger pore zeolites, this side reaction is greatly reduced with smaller pore ZSM-5 zeolite. As shown in Figure 6.4, the relative rate constant of the disproportionation reaction to that of the isomerization reaction decreases with the decrease in the pore size of the zeolite, suggesting that the pore size of the zeolite is the predominant factor influencing selectivity.

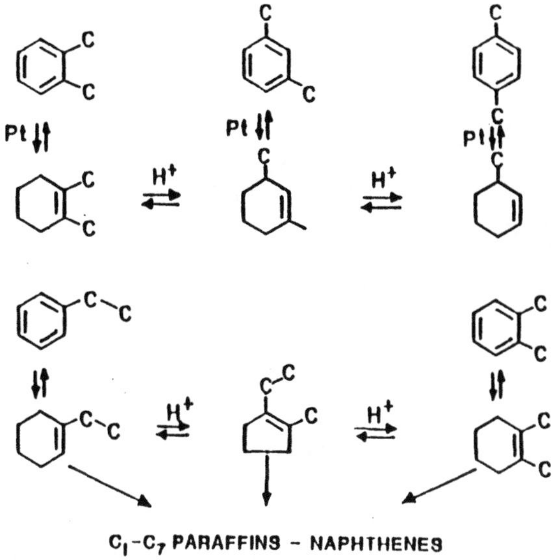

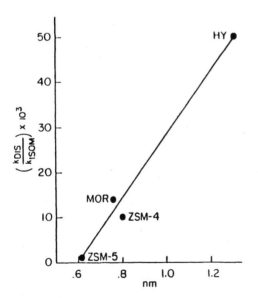

Pt
H+
C1-C7 PARAFFINS - NAPHTHENES

Figure 6.3 Bifunctional isomerization reactions—octafining. (From Olson and Haag, 1984.)

HY
MOR
ZSM-4
ZSM-5
nm

Figure 6.4 Relative rate constants of disproportionation to isomerization of xylenes over various zeolites. (From Olson and Haag, 1974.)

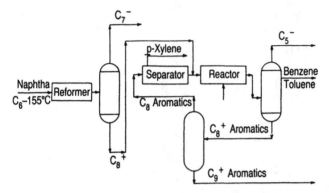

Figure 6.5 High-temperature xylene isomerization process. (From Tabak and Morrison, 1980.)

With ZSM-5, due to the combined effect of configurational diffusion and restricted transition state selectivity, the rate of transalkylation of ethylbenzene is more than 100 times faster than that of disproportionation of xylenes. Thus, the loss of xylenes is minimized—a major reason for its superb product selectivity in addition to its high activity and stability (Haag and Dwyer, 1979; Olson and Haag, 1984).

For unextracted feeds which may contain C_8^+ aliphatics, Mobil developed a high-temperature xylene isomerization process (MHTI). The catalyst contains a selective hydrogenation function and the process operates at higher temperatures (Olson and Haag, 1984). At these higher temperatures the aliphatics are cracked to lighter products, and the ethylbenzene is hydrodealkylated to benzene and ethane. Again xylene losses are exceedingly low. Figure 6.5 shows a schematic diagram of the process.

In 1990 a new high activity catalyst was developed for the MHAI process. Table 6.3 compares the product distribution of MHTI and MHAI at constant ethylbenzene conversion.

Both processes have catalyst cycle length of several years in commercial service. These new xylene isomerization processes are now in commercial use for more than half of the western world capacity for *p*-xylene production.

C. Xylene Synthesis

The Mobile Toluene Disproportionation Process (MTDP) is designed to convert toluene to a mixture of xylene isomers and benzene. The rate of toluene disproportionation is about 5000 times slower than that of xylene isomerization

Table 6.3 Comparison of Product Distribution for MHTI and MHAI
at Constant Ethylbenzene (EB) Conversion

	Process conditions: Extracted feed 15 atm, 10 WHSV, 2 H_2/HC			
	EB conversion, wt %			
	55		75	
Catalyst	MHTI	MHAI	MHTI	MHAI
Temperature, °C	448	412	475	436
Products, wt %				
C_5^-	1.42	1.32	2.00	2.01
Benzene	3.41	3.29	4.66	4.50
Toluene	2.39	2.29	3.09	2.14
Ethylbenzene	3.98	3.97	2.21	2.21
p-Xylene	20.96	21.17	20.70	20.88
m-Xylene	46.13	46.71	45.26	45.85
o-Xylene	20.56	20.37	20.63	20.38
C_9 Aromatics	0.93	0.68	1.34	0.98
C_{10} Aromatics	0.27	0.23	0.21	0.14
Xylene loss, wt %	2.15	1.55	3.25	2.81

Source: Huang et al. (1991).

(Olson and Haag, 1984). However, at above 450°C, toluene disproportion-
ates readily over ZSM-5 to give close to an equilibrium mixture of benzene
and xylenes.

The process employs a fixed-bed reactor operating in hydrogen at a total
pressure of 20 to 40 atm. Cycle life depends on the ratio of hydrogen to hydrocarbon. At a mole ratio of hydrogen to hydrocarbon of 1.5 to 3, a cycle life of 6
months to one year can be expected. Table 6.4 shows a typical operating result.

It was noted in the laboratory and in commercial operation that as the catalyst ages toward the end of cycle, it often undergoes subtle changes to exhibit
selective production of *p*-xylene. As shown by the data obtained in a laboratory
aging test (Figure 6.6), the *p*-xylene concentration slowly increased with on-
stream time and eventually exceeded its equilibrium value, reminiscent of the
para selectivity achieved by chemical modification (Chapter 4).

A new toluene disproportionation catalyst which operates at 50°C lower
temperatures while giving similar product selectivities has been described by
Absil et al. (1988).

Table 6.4   Toluene Disproportionation

Operating conditions:	
Pressure: 33 atm	
Temperature: 505°C	
H_2/toluene: 2.5 mol/mol	
LHSV: 6 vol/vol-h	
Toluene conversion, wt %:	42.6
Product distribution, wt %	
C_5^-	1.5
Benzene	17.9
Toluene	57.4
p-Xylene	4.7
m-Xylene	11.5
o-Xylene	5.2
Ethylbenzene	0.2
C_9^+	1.8

The process has been in commercial use at Mobil's Naples petrochemical complex in Italy since 1975. Three additional units have been licensed worldwide.

Competing with MTDP are the UOP Tatory and Arco Xylene Plus processes. However, they are all limited in terms of the *p*-xylene content in the

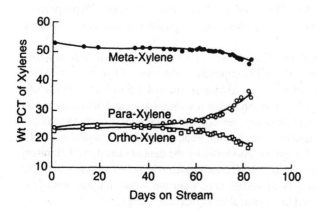

Figure 6.6   Xylene distribution in toluene disproportionation. (From Nicoletti, personal communication.)

mixed xylenes product by thermodynamic equilibrium (24% para isomer) and thus require a large recycle loop (i.e., separation followed by isomerization) to convert the other isomers to the para isomer.

To overcome this problem a selective toluene disproportionation process (MSTDP) (Olson and Haag, 1984), using a selectivated catalyst was developed by Mobil that can produce *p*-xylene and benzene in one step from toluene. The first commercial MSTDP operation was constructed in Italy in 1988 (Calvert et al., 1990).

D. Xylene by Alkylation of Toluene with Methanol

In a manner similar to toluene disproportionation, unmodified zeolite gives an equilibrium mixture of xylenes, while modified or selectivated ZSM-5 (Kaeding et al., 1984) gives predominantly the desired para isomer (Table 6.5). This provides another new route for the production of *p*-xylene where it is the major product of the reaction.

Table 6.5 Unselective and Selective Alkylation of Toluene with Methanol to Produce Xylenes and Water

	Catalyst		
	HZSM-5 (unselective)	Mg-ZSM-5 (selective)	P-ZSM-5 (selective)
Temperature, °C	495	600	600
Pressure	atm	atm	atm
Toluene/MeOH, mol ratio	2/1	1/1	1.5/1
Conversion, wt %			
Toluene	40	29	40
MeOH	99	99	96
Liquid effluent, wt %			
Toluene	56	67	57
Xylene	34	28	36
Other	10	5	7
Xylene isomers, %			
Para	24	86	90
Meta	53	10	7
Ortho	23	4	3

Source: Kaeding et al. (1984).

II. ETHYLBENZENE AND *PARA*-ETHYLTOLUENE

A. Ethylbenzene Synthesis

The Mobil/Badger Ethylbenzene process produces ethylbenzene, the feedstock for styrene, by alkylating benzene with ethene over a ZSM-5 catalyst. This is the first commercial alkylation process using an acid zeolite catalyst. The process has been in commercial use since 1976 (Dwyer, 1981). American Hoechst built their first Mobil/Badger Ethylbenzene/Styrene plant in Bayport, TX, in 1980. As of 1992, more than 19 commercial plants are operating around the world including the United States, Taiwan, Japan, Canada, Saudi Arabia, Korea, The Netherlands, the United Kingdom, and others, with a total annual ethylbenzene capacity of more than 5,700,000 metric tons (Dwyer and Ram, 1991).

A simplified process flow diagram is shown in Figure 6.7. In this process, fresh and recycled benzene is combined with a diethylbenzene-rich stream recovered from the product recovery section, fed to the alkylation reactor, and reacted with fresh ethene over ZSM-5 to give ethylbenzene. The mole ratio of benzene to ethene in the feed falls in the range of 5 to 20. The reaction takes place above 370°C, at 14–27 atm and at a very high weight hourly space velocity of 300–400 kg total feed/kg catalyst/hour. As the mole ratio of benzene to ethene is decreased, single-pass conversion of benzene increases at the expense of producing more diethylbenzene, which must be recovered and recycled.

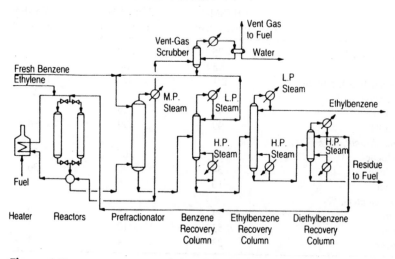

Figure 6.7 New catalyst simplifies Mobil/Badger for making ethylbenzene. (From Dwyer et al., 1976.)

The process has an overall yield of 99.6% ethylbenzene. Heavier residue production has been less than 0.3 wt % relative to ethylbenzene produced (Dwyer et al., 1976; Lewis and Dwyer, 1977).

Because of the low coke forming properties of ZSM-5, cycle times of up to 40–60 days between regenerations have been obtained. The high-temperature vapor phase alkylation process has major advantages over the low-temperature liquid–gas phase process in its energy recovery efficiency and process design simplicity, in addition to the elimination of waste disposal problems associated with the Friedel–Crafts type catalysts. The process can utilize dilute ethene feed streams.

In 1986 Badger and Mobile developed a new generation reactor system to further improve cycle length and yields (Dwyer and Ram, 1991). This new reactor system will reduce selectivity loss to byproducts from the earlier design by 50% and increase the reactor cycle length to over one year, which eliminates the need for a spare reactor. The new system was licensed by Taiwan Styrene Monomer Corporation and the plant was in operation successfully in March 1990.

B. *Para*-Ethyltoluene Synthesis

Mobil's *p*-Ethyltoluene (MPET) process is one of the new processes utilizing the unique configurational diffusion effect of the channel structure of ZSM-5 to selectively produce *para*-dialkylbenzenes (Kaeding et al., 1981). In the MPET process, *p*-ethyltoluene is produced by alkylating toluene with ethene over a chemically modified ZSM-5 (Kaeding et al., 1982). The product is 97% *p*-ethyltoluene (Table 6.6). Subsequent dehydrogenation of *p*-ethyltoluene gives the corresponding *p*-methylstyrene A multimillion-pound-per-month demonstration plant was in operation to provide material for product, process, and market development. The unique and desirable properties of poly-*para*-methyl styrene starting with toluene, effecting savings in aromatic raw material costs compared to benzene (the starting material for styrene), provide the potential for the establishment of a major new monomer in the marketplace (Kaeding et al., 1984).

III. SPIN-OFF TO OTHER CHEMICALS

The selectivity and stability advantages of zeolite catalysts may also be realized in the production of heterocompounds such as alcohols, ethers, and glycols. A cursory look at the patent literature shows that, for example, the double bond of unsaturated aldehydes and ketones can be isomerized over low acidity

Table 6.6 Alkylation of Toluene with
Ethyltoluene over Modified ZSM-5 Catalysts

Product composition, wt %	
Light gas and benzene	0.9
Toluene	86.2
Ethylbenzene and xylenes	0.5
p-Ethyltoluene	11.9
m-Ethyltoluene	0.4
o-Ethyltoluene	0
$C_{10}{}^{+}$ aromatics	0.1
Total	100.0
Ethyltoluene isomers, %	
para	96.7
meta	3.3
ortho	0
Total	100.0

Source: Kaeding et al. (1982).

medium pore zeolites without cleaving the carbonyl group, and aldehydes and ketones can be interconverted by skeletal isomerization.

In spite of the fact that acid sites are prone to be poisoned by basic nitrogen compounds and phenols, nitrogen compounds such as amines, pyridines, anilines, nitriles, and phenols have been synthesized with medium pore zeolites with the desired selectivity.

Aromatic alkylation reactions which led to a number of para selective products have been duplicated on phenols and thiophenes, as have halogenation and nitration reactions. Similar applications in the fine chemicals and pharmaceutical industries can also be expected. Curiously, development in these two areas has remained fairly dormant, at least to the outside world. Titanium-containing zeolite, TS-1, a novel derivative of ZSM-5, was commercialized at ENI of Italy Notari (1987) in the production of hydroquinone, catechol, and others.

REFERENCES

Abdul Hamid, S. B., E. G. Devouane, G. Demortier, J. Riga, and M. A. Yarmo, Appl. Catal. *108*, 83 (1994).

Absil, R. P. L., S. Han, S. M. Leiby, D. O. Marler, J. P. McWilliams, and D. S. Shihabi, "Toluene Disproportionation over ZSM-5 Catalyst," paper presented at the AIChE Summer National Meeting, Denver, CO, Aug. 21–24, 1988.

Anderson, R. F., J. A. Johnson, and J. R. Mowry, "Cyclar," paper presented at the AIChE Spring National Meeting, Houston, TX, Mar. 24–28, 1985.

Calvert, R. B., J. R. Green, S. A. Tabak, and S. Yurchak, "Shape Selective Catalysis— A Glimpse at the Future," paper presented at the Japan Petroleum Institute, Petroleum Refining Conference, Tokyo, Oct. 18–19, 1990.

Chen, N. Y. and W. E. Garwood, J. Catal. *52*, 453 (1978).

Chen, N. Y. and T. Y. Yan, Ind. Eng. Chem., Process Design Devel. *25*, 151 (1986).

Davies, E. E. and A. J. Kolombos, U.S. Pat. 4,180,689, Dec. 25, 1979.

Detz, C. M. and L. A. Field, U.S. Pat. 4,347,394, Aug. 31, 1982.

Dwyer, F. G., P. J. Lewis, and F. M. Schneider, Chem. Eng. *83(1)*, 90 (1976).

Dwyer, F. G., *Catalysis of Organic Reactions*, W. R. Moser, ed., Marcel Dekker, New York, p. 39, 1981.

Dwyer, F. G. and S. Ram, "Development and Commercialization of the Mobil/Badger Ethylbenzene Process," paper presented at the AIChE Spring National Meeting, Houston, TX, Apr. 7–11, 1991.

Euro. Chem. News, Feb. 24, 1992.

Garska, D. C. and B. M. Lok, U.S. Pat. 4,499,315-6, Feb. 12, 1985.

Haag, W. O. and D. H. Olson, U.S. Pat. 3,856,871, Dec. 24, 1974.

Haag, W. O. and F. G. Dwyer, "Aromatics Processing with Intermediate Pore Size Zeolite Catalysts," paper presented at the AIChE 8th National Meeting, Boston, MA, Aug. 19, 1979.

Huang, Y. Y., L. L. Breckenridge, R. A. Sailor, and W. O. Haag, "A Dual ZSM-5 Catalyst System for Xylene Isomerization," paper presented at the AIChE 1991 Spring National Meeting, Houston, TX (paper No. 30d), Apr. 7–11, 1991.

Johnson, J. A. and G. K. Hilder, "Dehydrocyclodimerization, Converting LPG to Aromatics," paper presented at NPRA Annual Meeting, San Antonio, TX, March 25–27, 1984.

Kaeding, W. W., C. Chu, L. B. Young, B. Weinstein, and S. A. Butter, J. Catal. *67*, 159 (1981).

Kaeding, W. W., L. B. Young, and A. G. Prapas, Chemtech *12*, 556 (1982).

Kaeding, W. W., G. C. Barlie, and M. M. Wu, Catal. Rev.-Sci. Eng. *26*, 597 (1984)

Kitagawa, H., Y. Sendoda, and Y. Ono, J. Catal. *101*, 12 (1986).

Kwak, B. S., W. M. H. Sachtler, and W. O. Haag, J. Catal. *149*, 465 (1994).

Lewis, P. J. and F. G. Dwyer, Am. Chem. Soc. 173rd Annual Meeting, New Orleans, Mar. 1977; Oil Gas J. *75(40)*, 55 (1977).

Mole, T. and J. R. Anderson, Appl. Catal. *17*, 141 (1985).

Notari, B., Stud. Surf. Sci. Catal. *37*, 413 (1987).

Oil Gas Journal *36*, 31 (1987).

Olson, D. H. and W. O. Haag, Am. Chem. Soc. Symp. Ser., *248*, 275 (1984).

Price, G. L. and V. Kanazirev, J. Catal. *126*, 267 (1990).

Tabak, S. A. and R. A. Morrison, U.S. Pat. 4,188,282, Feb. 12. 1980.

Walsh, D. E. and L. D. Rollmann, J. Catal. *56*, 195 (1979).

7

Applications in Alternate Fuels and Light Olefins

I. METHANOL-TO-GASOLINE PROCESS

The first methanol-to-gasoline (MTG) process was brought on stream in 1985 in New Zealand, producing at full capacity 14,500 barrels of gasoline per day. The gasoline produced from methanol contains typically about 32% aromatics, 60% saturates, and 8% olefins and is virtually identical to conventional unleaded gasoline of high research octane number (about 93).

Mobil's MTG process has attracted worldwide attention (Kam et al., 1984). Its discovery has been hailed as the first new route in more than 40 years for the production of gasoline from coal (Meisel, 1981).

The commercial development of this process has been centered on two versions of reactor design, fixed-bed and fluid-bed (Liederman et al., 1980, 1982). The chemistry and the operating characteristics of the conversion process were reviewed by Chang (1983). In designing these two types of reactors, the major concern is in the control and dissipation of the heat generated by the exothermic conversion reactions.

The first commercial fixed-bed MTG unit, a 14,500 barrel per day (B/D) gasoline unit, was erected in New Zealand to convert natural gas to gasoline via methanol synthesis. It has been on stream since late 1985. A 100 B/D fluid-bed demonstration plant was built at the Union Rheinishe Braunkohlen Kraftstoff AG (URBK) facility in Wesseling, near Bonn, West Germany in a joint study between Mobil and URBK and UHDE GmbH, both of Germany, with governmental financial support (Kam et al., 1984). The unit represents a horizontal scale-up of the 4 B/D unit. It has been in operation since 1982. Testing of the fluid-bed design was completed in 1984.

A. Commercial Fixed-Bed Plant

Shown in Figure 7.1 is the process flow diagram of the New Zealand plant. The crude methanol from the methanol synthesis unit is fed at about 21 atm to the MTG unit, which has four main sections: (1) reactors, (2) product separation by distillation, (3) heavy gasoline treaters to reduce durene concentration, and (4) catalyst regeneration facilities. The MTG reactors comprise a dehydration reactor followed by a parallel train of five conversion reactors. In the dehydration reactor, methanol is converted over a dehydration catalyst to an equilibrium mixture of methanol/dimethyl ether/water, releasing about 20% of the total reaction heat and raising the temperature of the stream from about 300 to about 400°C; this mixture is then mixed with 9:1 mol ratio of recycle gas and converted over ZSM-5 to hydrocarbons. The dilution by recycle gas controls the reactor exit temperature to 400–420°C. Heat is recovered from the effluent stream in a medium pressure stream generator and heat exchanger (Lee et al., 1980; Penick et al., 1982; Yurchak, 1988).

In actual operation, four of the five conversion reactors are on stream; the fifth reactor is either in the regeneration mode or on standby (Fox, 1982). During each operating cycle, the yield of C_5^+ gasoline slowly increases, and its composition slowly changes from an aromatic rich gasoline to an olefin-rich gasoline until methanol breakthrough, at which time the cycle ends and the catalyst oxidatively regenerated. Typical pilot plant data on the change in product composition with time on stream are shown in Table 7.1. Also shown in Table 7.1 is the yield of 9 RVP (Reid vapor pressure) gasoline, which includes the alkylate made from the isobutane and light olefinic products. With four parallel reactors in staggered operation, the composition of the total gasoline product is maintained constant with time.

It is unique to medium pore zeolites, through which various tetramethylbenzene isomers diffuse at vastly different rates, that durene (1,2,4,5-tetramethylbenzene), the smallest among them, is preferentially formed far above its equilibrium value. Shown in Table 7.2 is the durene content in the aromatic fraction of a typical gasoline product from the methanol reactor.

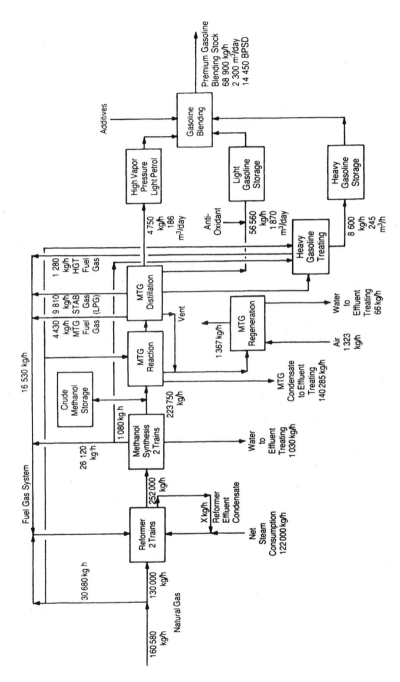

Figure 7.1 Process flow diagram of New Zealand MTG plant. (From Walker, 1985.)

Table 7.1   A Fixed-Bed Aging Study: Hydrocarbon Product Distribution

	Time on stream, h				
	8	47	95	142	167
Methane, ethane, ethylene	2.4	2.0	2.1	1.5	2.0
Propane	11.0	7.7	7.6	5.6	4.9
n-Butane	5.2	4.0	3.9	2.9	2.6
Isobutane	10.2	9.5	9.5	8.7	8.6
C_3–C_4 olefins	2.5	2.7	3.4	4.2	3.7
C_5^+ nonaromatics	35.0	42.1	42.4	48.3	50.8
Aromatics	33.7	32.0	31.1	28.8	27.4
	100.0	100.0	100.0	100.0	100.0
C_5^+	68.7	74.1	73.5	77.1	78.2
9 RVP gasoline including alkylate	75.9	81.9	82.8	88.2	88.4
Durene	2.5	4.2	4.3	4.2	4.5

Basis: weight percent.
Source: Chang (1983).

Table 7.2   Durene Content in Aromatics from Methanol

% A_{10} in aromatics: 8.0	% Durene in A_{10}: 50.1
	Equilibrium at 370°C: 33.4

Source: Chang et al. (1981).

Because of its relatively high melting point (79°C), durene at sufficiently high concentrations could crystallize to form a deposit in the carburetor during cold starts. However, extensive vehicle performance tests (Fitch and Lee, 1981) showed no adverse effect due to durene when its concentration is less than 2%. Conventional petroleum-derived gasolines in the United States typically contain 0.2 to 0.5% durene.

To reduce the durene concentration in the MTG product, a heavy gasoline treating process, typical of a mild hydrotreating process, was developed to process the 177°C+ bottom fraction, which contains all the durene (Fox, 1982; Penick et al., 1982). As shown by the data in Table 7.3, the concentration of durene in the heavy gasoline is reduced by isomerization, dealkylation and hydrocracking to less than 15 wt % (Yurchak, 1988) without suffering any loss of

Table 7.3 Heavy Gasoline Treating Yields (wt %)

	Feed	Product
C_2^-	—	0.3
$C_3 + C_4$	—	1.96
C_5^+ P + N	0.65	4.52
C_6 aromatics	0.00	0.03
C_7 aromatics	0.00	0.52
C_8 aromatics	0.74	5.74
C_9 aromatics	23.04	29.41
Durene, 1,2,4,5-tetramethylbenzene	43.69	13.62
Isodurene, 1,2,3,5-tetramethylbenzene	8.13	16.06
Prehnitene, 1,2,3,4-tetramethylbenzene	2.80	3.60
Other C_{10}^+ aromatics	20.95	24.24

Source: Yurchak (1988).

Table 7.4 Melting Point of Tetramethyl Benzenes

	Melting point, °C
Prehnitene, 1,2,3,4-tetramethylbenzene	−23.7
Isodurene, 1,2,3,5-tetramethylbenzene	−6.3
Durene, 1,2,4,5-tetramethylbenzene	+79.2

Source: Rossini et al. (1953).

octane rating. It is interesting to note that among the isomers only durene has a high melting point (Table 7.4).

The finished gasoline product from a fixed-bed MTG process comprises a mixture of MTG gasoline and varying quantities of butanes and alkylates, depending on volatility requirements and on the availability of an alkylation unit. The composition and properties of a typical finished MTG gasoline are shown in Table 7.5.

B. Fluid-Bed Development

Development of the fluid bed process began with the scale-up of a bench unit to a 4 B/D pilot plant (Liederman et al., 1978; Kam and Lee, 1978). Fluidized

Table 7.5   Typical Finished Gasoline Properties

Components, vol %	
C_5^+ Gasoline	95
C_4's	5
Octane number	
Clear research	93
Clear motor	83
Distillation (D-86), RC (RF)	
10%	46 (115)
50%	99 (210)
90%	166 (330)
EP	204 (400)
Existent gum, mg/100 ml	1
Sulfur, nitrogen, ppm	<10
Composition, wt %	
n-Paraffins	4.6
i-Paraffins	41.6
Olefins	9.5
Naphthenes	9.2
Aromatics	35.1
Durene	1.84

Source: Yurchak (1988).

beds offer the advantage over fixed beds in the ease of temperature control and heat removal, but are complicated by the necessity that the internals of the reactor must be designed to achieve complete conversion of methanol in order to avoid the costly step of recovering unconverted methanol from the reactor effluent. The fluid-bed unit operates at much lower pressures than the fixed-bed unit, and its yield pattern is different, having lower C_5^+ liquid yield and higher light olefins than yield of fixed-bed units. Table 7.6 compares the process conditions and product yields from these two systems (Penick et al., 1982). The composition and properties of a typical finished gasoline from a fluid-bed pilot plant are shown in Table 7.7. Performance of the bench unit and the 4 B/D pilot plant was recently reviewed by Chang (1983).

A schematic diagram of the 100 B/D demonstration plant is shown in Figure 7.2. The reactor consists of a dense fluid-bed section located above a dilute phase riser. Reaction heat is removed either by circulating the cata-

Table 7.6   Process Conditions and Product Yields from MTG Processes

	Fixed bed	Fluid bed
Conditions		
Methanol/water chg. (W/W)	83/17	82/17
Dehydration reactor inlet temperature (°C)	316	
Dehydration reactor outlet temperature (°C)	304	—
Conversion reactor inlet temperature (°C)	360	413
Conversion reactor outlet temperature (°C)	415	413
Pressure (kPa)	2,170	275
Recycle ratio (mol/mol change)	9.0	—
Space velocity (WHSV)	2.0	1.0
Yields, wt % of methanol charged		
Methanol + ether	0.0	0.2
Hydrocarbons	43.4	43.5
Water	56.0	56.0
CO, CO_2	0.2	0.2
	100.0	100.0
Hydrocarbon product, wt %		
Light gas	1.4	5.6
Propane	5.5	5.9
Propylene	0.2	5.0
Isobutane	8.6	14.5
n-Butane	3.3	1.7
Butenes	1.1	7.3
C_5^+ gasoline	79.9	60.0
	100.0	100.0
Gasoline (including alkylate) [RVP-62 kPa (9 psi)]	85.0	88.0
LPG	13.6	6.4
Fuel gas	1.4	5.6
	100.0	100.0

Source: Penick et al. (1982).

lyst through an external cooler or through heat exchangers placed within the catalyst bed to produce steam at pressures up to 100 atm (Penick et al., 1982; Gierlich et al., 1985). Table 7.8 lists the operating variables of the fluid-bed demonstration plant.

    One distinct advantage of the fluid-bed over the fixed-bed reactor is its ability to maintain constant catalyst activity by continuous regeneration

Table 7.7 Typical Properties of Finished Gasoline from Fluid-Bed MTG Unit

Components, wt %		
Butanes	3.2	
Alkylates	28.6	
C_5^+ gasoline	68.2	
	100.0	
Composition, wt %		
Paraffins	56	
Olefins	7	
Naphthenes	4	
Aromatics	33	
	100	

Octane	Research	Motor
Clear	96.8	87.4
Reid vapor pressure	9	
Specific gravity	0.730	
Sulfur, wt %	Nil	
Nitrogen, wt %	Nil	
Durene, wt %	3.8	
Corrosion, copper strip	1A	

ASTM Distillation, RC	
10%	47
30%	70
50%	103
90%	169

Source: Lee et al. (1979).

and thereby produce a constant quality product. By adjusting the operating variables, i.e., temperature and pressure, it is possible to maximize either the gasoline yield or its octane rating, as shown by the data in Figures 7.3 and 7.4.

Data obtained in the demonstration plant will serve as the basis for a conceptual design of a commercial size fluid-bed reactor, processing 2,500 tons/day (18,000 B/D) of methanol (Gierlich et al., 1985). The design was completed in 1985 and is available for commercial license.

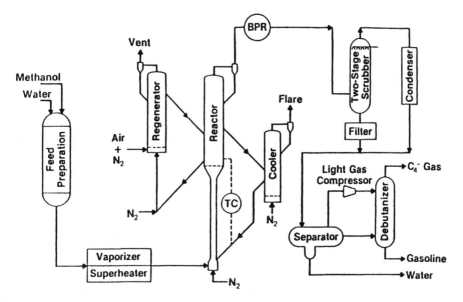

Figure 7.2 100 B/D fluid MTG plant. (From Penick et al., 1982.)

Table 7.8 Operating Variables of Fluid-Bed Demonstration Plant

Reactor pressure	2.7–4.5 bars abs.
Reactor temperature	380–430°C
WHSV	0.5–1.75 H^{-1}
Methanol feed rate	500–1050 kg/h
Methanol conversion	>99.9%
Time on stream	8600 h
Methanol processed	6800 metric tons
Max. gasoline yield (incl. alkylate)	92%
Service factor total	65%
Service factor during scheduled test runs	99%

Source: Gierlich et al. (1985).

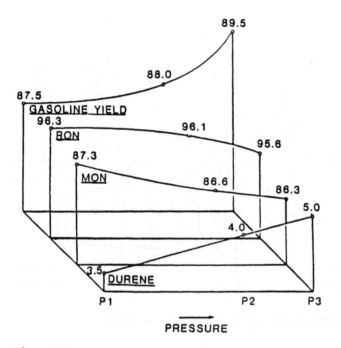

89.5

88.0

87.5

GASOLINE YIELD

96.3

RON

96.1

95.6

87.3

MON

86.6

86.3

5.0

4.0

3.5 DURENE

P1   P2   P3

PRESSURE

Figure 7.3   Effect of pressure on gasoline quality. (From Gierlich et al., 1985.)

II. METHANOL-TO-LIGHT OLEFINS PROCESS

Instead of producing an aromatic gasoline, methanol can also be converted to yield mostly light olefins. Mobil's Methanol-to-Olefins (MTO) process is based on a discovery that by reducing the acidity of the zeolite and raising the operating temperature to above 500°C, the relative rate of olefin formation reactions and aromatization reactions are different enough so that C_2 to C_5 olefin yield as high as 80% has been obtained (Chang et al., 1984b). These laboratory results have been confirmed in a 4 B/D fluid bed pilot plant (Gould et al., 1986). Shown in Table 7.9 is a comparison of the results obtained in these two units.

The process has also been successfully demonstrated in the 100 barrel per day fluid-bed MTG unit in Germany (Gould et al., 1986). The 100 B/D reactor is 60 cm in diameter and 12 m high, a modified version of the MTG reactor. The reactor is equipped with heat exchanger coils immersed in the dense fluid bed. Because the unit was originally designed for the MTG process, the reactor could not be operated at near atmospheric pressure, hence the demonstration runs were carried out at 2.5 atm and 500°C. As expected, higher operating

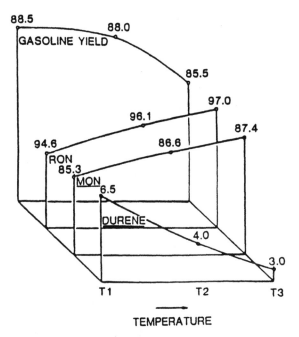

Figure 7.4 Effect of temperature on gasoline quality. (From Gierlich et al., 1985.)

pressure reduces the olefin yield to 61% and increases the gasoline yield to 32% (Table 7.10). Considering the excellent reproducibility between the pilot plant and the demonstration unit, the scale-up study was considered a success.

The MTO process may be coupled with Mobil's Olefins-to-Gasoline and Distillate (MOGD) process described earlier to provide another synthetic fuel process for producing gasoline and distillates from methanol and/or coal. A schematic diagram of the combined process is shown in Figure 7.5 (Tabak et al., 1986). Methanol is fed to the MTO reactor where it is converted to hydrocarbons and water. The hydrocarbons are separated into light olefins, an ethene-rich stream and an aromatic gasoline stream, which is blended into the final gasoline product. The light olefins are fed to the MOGD unit along with an MOGD recycle stream. The MOGD product is then separated into raw distillate, gasoline, and small fuel gas and LPG streams. Hydrotreating the raw distillate gives the final distillate product.

When the products of the MTO/MOGD processes are compared with that of Fischer-Tropsch synthesis, it is noted that in the MTO process, the yield of C_3^- paraffins is typically less than 5 wt % of total hydrocarbons. This is

Table 7.9 Comparison of Bench Scale and Pilot Plant Data on Methanol-to-Olefin Process

Reaction conditions: Pressure: 1 atm, temperature: 482°C

	wt %	
	Bench unit	Pilot plant
Methanol conversion	100	100
Product yields		
Ethene	5.8	5.2
Propene	33.7	32.9
Butenes	18.4	19.1
Pentenes	14.8	12.0
C_6^+ olefins	5.5	7.7
C_3^- paraffins	5.0	5.0
C_4^+ paraffins/naphthenes	10.3	11.6
Aromatics	6.5	6.5
	100.0	100.0
Total olefins	78.2	76.9
Total C_4^+ gasoline	16.8	18.1
Light paraffins	5.0	5.0

Source: Gould et al. (1986).

Table 7.10 Comparison of Pilot Plant and Demonstration Unit Data on Methanol-to-Olefin Process

Reaction conditions: Pressure: 2.5 atm, temperature: 500°C

	wt %	
	Pilot plant	Demonstration unit
Methanol conversion	100	100
Product yields		
C_3^- paraffins	7	7
Olefins	63	61
Total C_4^+ gasoline	30	32
	100	100

Source: Gould et al. (1986).

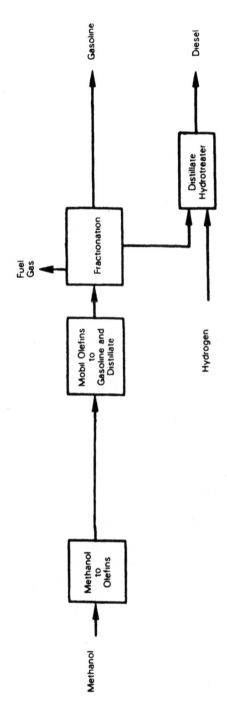

Figure 7.5 An example of an integrated MTO/MOGD process (schematic). (From Tabak et al., 1986.)

Table 7.11   MTO/MOGD Product Yields

Distillate/gasoline ratio	0.65	0.95
Product yield, wt %		
LPG	5.1	5.1
Gasoline	57.0	48.0
Diesel	37.9	46.9
	100	100

Source: Tabak et al. (1986).

significantly less than that produced by Fischer-Tropsch synthesis (typically more than 15%). Thus it is possible to produce liquid yields in excess of 90% of the total hydrocarbon products by this new route.

Since both MTO and MOGD have flexible product slates, the combined process should be able to meet the variable seasonal market demand of gasoline or distillate. Table 7.11 presents an estimate of product yields from this combined process. It also provides a means to convert methanol to high quality jet and diesel fuels.

More recently, Union Carbide (1986) announced the development of a fluid-bed methanol-to-olefin process using a silico-aluminophosphate (SAPO) catalyst. The SAPO catalyst reportedly produces olefins in very high yields (>90%), and the process can be modified to produce 60% ethene or propene (Kaiser, 1985; Pellet et al., 1987; Pop et al., 1992).

III. SYNTHESIS OF ETHERS

As briefly described in Chapter 4, methyl *tert*-butyl ether (MTBE), an important commercial product as an oxygenate for gasoline blending, has been synthesized by reacting isobutene with methanol over Amberlyst 15, a microreticular hydronium ion–exchanged resin that can not tolerate high temperature regeneration. To overcome its shortcoming, some effort was made by Chu and Kühl (1987) to use the medium pore zeolite HZSM-5 or HZSM-11, which are regenerable at higher temperatures, as the catalyst. Their data on fixed bed reactors obtained at higher temperatures are impressively close to the equilibrium values; however, so far no commercial application has been attempted.

Ether such as diisopropyl ether (DIPE) can be synthesized by the hydration of propene (see Chapter 4). DIPE, if produced at a lower cost than MTBE, is an acceptable fuel oxygen source (McNally et al., 1992). In extensive laboratory, engine, and automobile fleet tests, DIPE performs equivalently to MTBE at blending levels up to 2.7 wt % fuel oxygen. Toxicology studies have shown no adverse health or environmental effects associated with exposure to gasoline blends containing MTBE.

To take advantage of the thermodynamics of the hydration reaction, the DIPE process operates at 70–140 atm pressure and lower than 260°C (Mao et al., 1993). The selectivity for the ether can be maintained at about 90%.

The DIPE process consists of three steps: feed preparation, reactor, and product recovery, as shown in Figure 7.6. The fresh propene/propane (PP) feed is treated in a water wash tower to remove trace contaminants. The PP splitter is included to increase the concentration of propene in the feed to the DIPE reactor, where it is mixed with the isopropanol (IPA)/water recycle stream. This recycle stream is pumped around the reactor to control the exothermic hydration reaction temperature. The effluent from the DIPE reactor is depressurized and cooled before it is sent to the PP stripper, where the unconverted propene and propane are recycled back to the PP splitter. The overhead from the PP

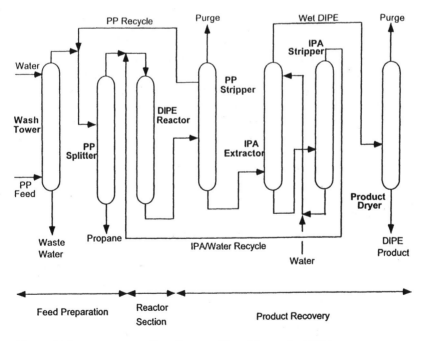

Figure 7.6 DIPE process flow diagram. (From Mao et al., 1993.)

stripper is purged to vent the lighter gas and the unreacted propane out of the system. The DIPE product, intermediate IPA, and water are sent to a separation unit where the IPA and water are recovered and recycled. Finally, the wet DIPE product is further dehydrated by distillation or solid bed adsorption.

IV. SYNTHESIS GAS CONVERSION

A. Composite Catalysts for Direct Synthesis Gas Conversion

While zeolites do not have sufficient intrinsic catalytic activity for CO reduction, they may be combined with carbon monoxide reduction catalysts such as methanol synthesis catalysts (Chang et al., 1984a; Fujimoto et al., 1984) or iron Fischer-Tropsch catalysts (Chang et al., 1979; Caesar et al., 1979; Dwyer and Garwood, 1984) and ruthenium Fischer-Tropsch catalyst (Huang and Haag, 1981a; Chen et al., 1984) to alter the distribution of the final product. Fischer-Tropsch synthesis typically produces a mixture of paraffins and olefins ranging from methane to high molecular weight waxes. The formation of C_5^+ hydrocarbons is usually favored by low reaction temperatures. By mixing the conventional Fischer-Tropsch catalyst with medium pore molecular sieves, such as HZSM-5 (Caesar et al., 1979) or SAPO-11 (Coughlin and Rabo, 1985; Coughlin, 1986), it has been demonstrated in the laboratory that the boiling range of the hydrocarbon product can be narrowed, and aromatic liquid products boiling in the gasoline range have been made in a single step directly from the synthesis gas. The olefin isomerization activity is also enhanced by the presence of the molecular sieve component to reduce the pour point of the liquid product. In the absence of the molecular sieve component, the liquid product comprises essentially waxy normal olefins and paraffins. The acidity of the SiO_2/Al_2O_3 ratio of the ZSM-5 in an intimate mixture of ZSM-5 and Fe(K) catalysts were found to have a significant effect on the olefinicity of the hydrocarbon products. Olefin contents increased with increasing SiO_2/Al_2O_3 ratios (Dwyer and Garwood, 1984) as shown by the data in Figure 7.7. The degree of intimacy of mixing of these two components in the composite catalyst was also found to affect the composition of the liquid product (Huang and Haag, 1981a). For example, Table 7.12 shows that without ZSM-5, the ruthenium Fischer-Tropsch catalyst makes only nonaromatic product. Adding ZSM-5 to the composite catalyst as a physical mixture changes the composition of the liquid product to that of an aromatic gasoline. Using a more intimate mixture, prepared by impregnating ZSM-5 with $RuCl_3 \cdot H_2O$, promotes the aromatics alkylation reaction and shifts the products to the C_{11}^+ heavy aromatics.

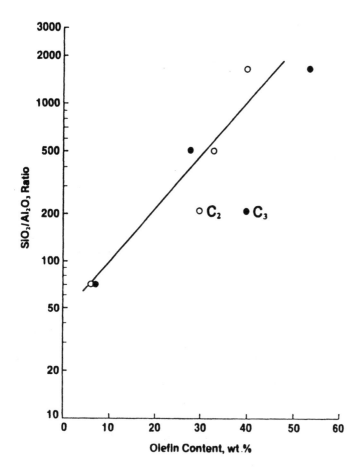

Figure 7.7 Effect of SiO_2/Al_2O_3 ratio on C_2 and C_3 olefin content. (From Dwyer and Garwood, 1984.)

One of the disadvantages in using these composite catalysts for the direct conversion of synthesis gas occurs in situations where maximizing C_5^+ product is desired. The problem lies in a mismatch of the reaction temperature required for optimal performance and is illustrated by the data shown in Figure 7.8, obtained with a physical mixture of RuO_2 and HZSM-5 over a temperature range of 260 to 320°C. Because the zeolitic component shows little catalytic activity below 260°C, while C_4^- gas increases rapidly with temperatures beyond 260°C with known Fischer-Tropsch catalysts, it is difficult to

Table 7.12   Effect of Intimacy of Mixing on Product Distribution

Reaction conditions: Pressure: 51 atm,
temperature: 294–304°C, H_2/CO: 2 mol/mol

		Catalyst	
	Ru/Al_2O_3	Ru/ZSM-5	Ru/Al_2O_3 + ZSM-5
Ru loading, wt %	0.5	5	5
Conversion, mol %	94	86	99
Reactor effluent, wt %			
Hydrocarbons	37	38	40
H_2	0	1	0
CO	11	15	0
CO_2	6	2	20
H_2O	46	44	40
Hydrocarbon composition, wt %			
$C_1 + C_2$	33	38	43
$C_3 + C_4$	8	16	14
C_5^+	59	46	43
Aromatics in C_5^+	0	25	24
Aromatics distribution			
Benzene-A_{10}	0	79	97
A_{11}^+	0	21	3

Source: Huang and Haag (1981a).

select a temperature at which the zeolitic component is effective without excess gas production.

B.   Mobil Two-Stage Slurry Fischer-Tropsch/ZSM-5 Process

Low methane formation and high aromatic gasoline production can be achieved when the Fischer-Tropsch catalyst and the zeolite catalyst are placed in separate reactors and operated in a cascade mode under different reaction conditions (Huang and Haag, 1981b).

This is the basis of Mobil's Two-Stage Slurry Fischer-Tropsch/ZSM-5 process currently under development. The process combines the slurry-phase Fischer-Tropsch synthesis technology with a fixed-bed ZSM-5 reactor, which converts the vapor phase product from the first-stage reactor to high quality gasoline (Kuo, 1983). To minimize the yield of methane and ethane, the operation mode of the slurry reactor leads to the production of C_{20}^+ waxy prod-

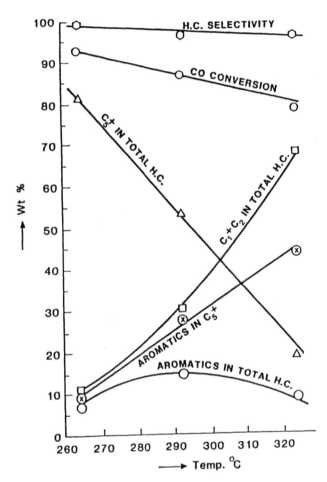

Figure 7.8   Effect of temperature on synthesis gas conversion over 5% Ru(as RuO₂)/ ZSM-5 (294°C, GHSV = 480, and H₂/CO = 2/1). (From Huang and Haag, 1981b.)

ucts which accumulate in the reactor and are removed for further upgrading (Kuo, 1985).

Figure 7.9 shows a simplified flow diagram of the two-stage pilot plant. The slurry Fischer-Tropsch reactor, shown in Figure 7.10, measures 5 cm inner diameter, and the liquid depth may be varied between 305 and 760 cm. It was designed to hold about 1600 g of catalyst and process up to 3 m³/h (STP) of synthesis gas. Reaction heat is removed from the reactor by circulating a

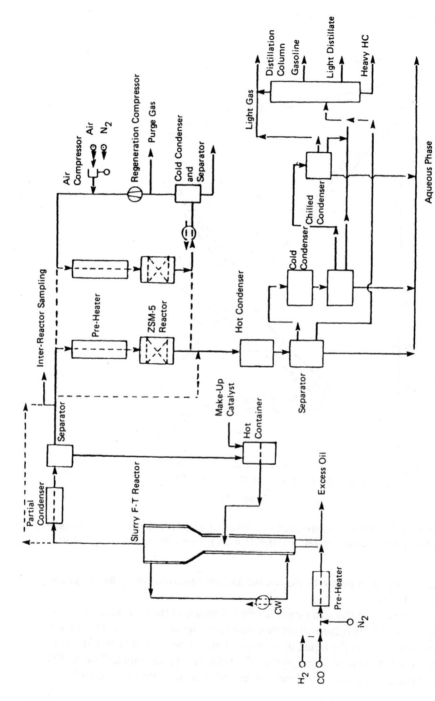

Figure 7.9 Simplified flow diagram of two-stage pilot plant for synthesis gas conversion. (From Kuo, 1983.)

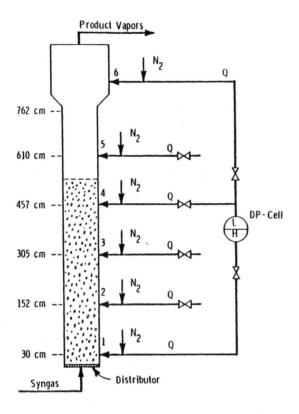

Q: DP-Cell Lines

Figure 7.10 Schematic arrangement of DP-cell for liquid level measurement. (From Kuo, 1983.)

hydrocarbon-based coolant through an outer jacket. The second stage comprises two ZSM-5 reactors. Each consists of a fixed catalyst bed, 5 cm in diameter and 20 cm long. One of them is in operation while the other is either in the regeneration mode or on standby.

Typical pilot plant data on the first-stage reactor feeding a syngas containing 0.68 mole ratio of hydrogen to CO are presented in Table 7.13. The overhead effluent from the reactor was processed in the second stage reactor to the final product (Table 7.14). Similar to the fixed-bed MTG reactor, during each operating cycle the catalyst slowly deactivated and the composition of

Table 7.13 Typical Pilot Plant Data on First-Stage Slurry Fischer-Tropsch Reactor

Operation conditions:	
Temperature: 260°C	
Pressure: 15 atm	
Space velocity: 2 1/g Fe-hr	
Catalyst: Fe/Cu/K$_2$CO$_3$	
Catalyst loading: 19.5 wt %	
Syngas conversion, mol %	
Hydrogen	78.8
CO	88.7
H$_2$ + CO	84.7
Yields, wt % of products	
Hydrocarbons (including oxygenates)	22.0
CO$_2$	66.4
H$_2$O	0.8
Hydrogen	0.9
CO	9.9
Composition of hydrocarbons, wt %	
Methane	7.9
Ethene	1.5
Ethane	3.0
Propene	8.2
Propane	1.9
Butenes	6.0
Butanes	2.0
C$_5$+ overhead	54.3
Slurry wax	6.2

Source: Kuo (1983).

C$_5$+ gasoline slowly changed from an aromatic rich gasoline to an olefin rich gasoline. To maintain constant quality product, the reactor temperature was gradually raised from the start of cycle temperature of 300°C to the end of cycle temperature of about 410°C over a period of 30 days, at which time the catalyst was oxidatively regenerated. The average gasoline contains about 45 wt % of aromatics, 15 wt % olefins, and 40 wt % paraffins and has a research clear octane (R+0) of about 90.

The slurry wax removed from the first-stage reactor contains catalyst fines. After settling to remove most of the catalyst, it is vacuum distilled to re-

Table 7.14 Typical Pilot Plant Data
on Second Stage ZSM-5 Reactor

Operation conditions: Temperature: 320°C Pressure: 13 atm	
Overall yields, wt %	
Methane	7.3
Ethene	1.0
Ethane	3.3
Propene	0.9
Propane	8.5
Butenes	1.4
Butanes	19.0
C_5–C_{11} Gasoline	49.3
C_{12}^+ and slurry wax	9.3
	100.0

Source: Kuo (1985).

cover a solids-free wax stream, and the still bottom is thermally cracked and recycled to the vacuum tower. The wax stream can be upgraded to gasoline and
distillates by conventional hydrocracking/cracking processes. Based on the pilot plant study, a conceptual design for a 27,000 B/D gasoline plant has been
developed (Kuo, 1985).

REFERENCES

Caesar, P. D., J. B. Brennan, W. E. Garwood, and J. Ciric, J. Catal. *56*, 274 (1979).

Chang, C. D., Catal. Rev.-Sci. Eng. *25*, 1 (1983); *Hydrocarbons from Methanol*, Marcel Dekker, New York, 1983.

Chang, C. D., W. H. Lang, and A. J. Silvestri, J. Catal. *56*, 268 (1979).

Chang, C. D., W. H. Lang, and W. K. Bell, *Catalysis Organic Reactions*, W. R. Moser, ed., Marcel Dekker, New York, p. 73, 1981.

Chang, C. D., C. T. -W. Chu, and R. F. Socha, J. Catal. *86*, 289 (1984a).

Chang, C. D., J. N. Miale, and R. F. Socha, J. Catal. *90*, 84 (1984b).

Chen, Y. W., H. T. Wang, and J. G. Goodwin, J. Catal. *85*, 499 (1984).

Chu, P. and G. H. Kühl, I&EC Res. *26*, 365 (1987).

Coughlin, P. K., U.S. Pat. 4,579,830, Apr. 1, 1986; U.S. Pat. 4,632,941, Dec. 12, 1986.

Coughlin, P. K. and J. A. Rabo, U.S. Pat. 4,556,645, Dec. 3, 1985.

Dwyer, F. G. and W. E. Garwood, "Catalytic Conversions of Synthesis Gas and Alcohols to Chemicals," R. G. Herman, ed., Plenum, New York, p. 167, 1984.

Fitch, F. B. and W. Lee, "Methanol-to-Gasoline, An Alternative Route to High-Quality Gasoline," paper presented at the International Pacific Conference on Automotive Engineering, SAE Tech. Paper 811403, Honolulu, Nov. 16–19, 1981.

Fox, J. M. "The Fixed-Bed Methanol-to-Gasoline Process Proposed for New Zealand," paper presented at the Coal Gasification Conference, Australian Inst. Petro., Adelaide, Australia, Mar. 2, 1982.

Fujimoto, K., K. Yoshihiro, and H. Tominaga, J. Catal. *87*, 101 (1984).

Gierlich, H. H., K. H. Keim, N. Thiagarajan, E. Nitschke, A. Y. Kam, and N. Daviduk, "Successful Scale-Up of the Third Bed Methanol to Gasoline (MTG) Process to 100 BPD Demonstration Plant," paper presented at the 2nd EPRI Conf. "Synthetic Fuels—Status and Directions," San Francisco, Apr. 15–19, 1985.

Gould, R. M., A. A. Avidan, J. L. Soto, C. D. Chang, and R. F. Socha, "Scale-Up of a Fluid-Bed Process for Production of Light Olefins from Methanol," paper presented at the AIChE National Meeting, New Orleans, April 6–10, 1986.

Huang, T. J. and W. O. Haag, ACS Symp. Ser. *152*, 307 (1981a).

Huang, T. J. and W. O. Haag, "Aromatic Gasoline Production from Synthesis Gas via a Two-Stage Process," paper presented at the 91st AIChE National Meeting, Detroit, Aug. 16–19, 1981b.

Kaiser, S. W., U.S. Pat. 4,499,327, Feb. 12, 1985; U.S. Pat. 4,524,234, Jun. 18, 1985.

Kam, A. Y. and W. Lee, "Fluid-Bed Process Studies on Selective Conversion of Methanol to High Octane Gasoline," Final Report, DOE Contract No. EX-76-C-01-2490, U.S. Department of Energy, Apr. 1978.

Kam, A. Y., M. Schreiner, and S. Yurchak, "Handbook of Synfuels Technology," R. A. Meyers, ed., McGraw-Hill, New York, pp. 2–75, 1984.

Kuo, J. C. W., "Slurry Fischer-Tropsch/Mobil Two-Stage Process of Converting Syngas to High Octane Gasoline," Final Report, DOE Contract No. DE-AC22-80PC30022, June 1983.

Kuo, J. C. W., "Two-Stage Process for Conversion of Synthesis Gas to High Quality Transportation Fuels," Final Report, DOE Contract No. DE-AC22-83PC60019, Oct. 1985.

Lee, W., J. Maziuk, V. W. Weekman, and S. Yurchak, *Large Chemical Plants*, G. F. Froment, ed., Elsevier, Amsterdam, p. 171, 1979.

Lee, W., S. Yurchak, D. Davidukm and J. Maziuk, "A Fixed-Bed Process for Methanol-to-Gasoline Conversion," Paper AM-80-41, NPRA Meeting, New Orleans, Mar. 22–24, 1980.

Liederman, D., S. M. Jacob, S. E. Voltz, and J. J. Wise, Ind. Eng. Chem., Process Des. Dev. *17*, 340 (1978).

Liederman, D., S. Yurchak, J. C. W. Kuo, and W. Lee, "Mobil Methanol-to-Gasoline Process," paper presented at the 15th Intersoc. Energy Conv. Eng. Conf., Seattle, Aug. 1980.

Liederman, D., S. Yurchak, J. C. W. Kuo, and W. Lee, J. Energy *6*, 340 (1982).

Mao, C.-H., J. L. Soto, J. A. Stoos, R. A. Ware, and S. Yurchak, "Mobil DIPE Process—An Alternate Source for Oxygenates," paper presented at the World Solid Acid Proc. Conf. Houston, Nov. 14–16, 1993.

McNally, M. J., P. J. Costello, B. E. Johnson, C. M. Sorensen, S. S. Wise, and L. L. Low, "Hydration of Propylene to Diisopropyl Ether—A New Source of Fuel Oxygen," paper presented at the 1992 Annual NPRA Meeting, New Orleans, (AM-92-28) Mar. 23–25, 1992.

Meisel, S. L., Philos. Trans. Roy. Soc. London *A300*, 157 (1981).

Pellet, R. J., J. A. Rabo, G. N. Long, and P. K. Coughlin, "Reactions of C_8 Aromatics Catalyzed by Aluminophosphate Based Molecular Sieves," paper No. B-2, 10th North American Meeting of the Catalysis Society, San Diego, May 17–22, 1987; R. J. Pellet, J. A. Rabo, and G. N. Long, "Olefin Oligomerization Catalyzed by Aluminophosphate Based Molecular Sieves," poster paper No. 75, 10th North American Meeting of the Catalysis Society, San Diego, May 17–22, 1987.

Penick, J. E., W. Lee, and J. Maziuk, "Development of the Methanol-to-Gasoline (MTG) Process," paper presented at the International Symposium of Chemical Reactor Engineers (ISCRE-7), Boston, October 4, 1982.

Pop, G., G. Musca, D. Ivanescu, E. Pop, G. Maria, E. Chirila, and O. Munteam, Chem. Ind. *46*, 443 (1992).

Rossini, F. D., K. S. Pitzer, R. L. Arnett, R. M. Braun, and G. C. Pimental, "Selected Values of Physical and Thermodynamics Properties of Hydrocarbons and Related Compounds," API Research Project 44, Carnegie Press, Pittsburgh, 1953.

Tabak, S. A., A. A. Avidan, and F. J. Krambeck, "Production of Synthetic Gasoline and Diesel Fuels from Non-Petroleum Sources," paper presented at the ACS National Meeting, New York, Apr. 13–15, 1986.

Union Carbide Corp., Oil Gas J. *84*(50), 31 (1986).

Walker, B. V., "Synthetic and Alternative Fuels Production in New Zealand," paper presented at EPRI Conference on Coal Gasification and Synthetic Fuels for Power generation, San Francisco, Apr. 14–19, 1985.

Yurchak, S., Stud. Surf. Sci. Catal. *36*, 251 (1988).

8

New Opportunities in Shape Selective Catalysis

I. UPGRADING WAXY CRUDES

In principle, shape selective cracking is applicable to the processing of waxy crudes to improve its fluidity and eliminate the need to install heated pipelines to transport these crudes. Laboratory experiments have demonstrated that paraffinic waxy crude oils from various parts of the world, including the United States (Utah), Indonesia, Australia, North Africa, China, and Newfoundland (Chen and Shihabi, 1981; Shihabi, 1981), can be processed in a simple fixed-bed reactor with or without the presence of hydrogen to improve their low temperature fluidity and at the same time increase the quantity and quality of their distillate fraction.

Figure 8.1 illustrates the upgrading of a barrel of Daqing (Taching) whole crude into low pour point products. The conversion reduces the heavy fuel oil fraction from 76% to 52% and decreases the pour point of that fraction to $-16°C$ (Wise et al., 1986).

Using an activated NaZSM-5 catalyst in aging studies running in the absence of hydrogen under a moderate pressure of 15 to 30 atm of light gases, the catalyst life exceeded over one month (Figure 8.2) (Shihabi, 1983). The deactivated catalyst activity could be restored by water. Operating in such a

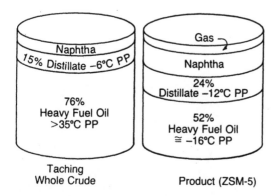

Figure 8.1 Opportunities for direct catalytic upgrading of waxy taching whole crude. (From Wise et al., 1986.)

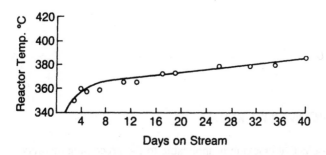

Figure 8.2 Daqing (taching) crude temperature-adjusted to −12°C pour point: LHSV = 1, pressure = 22 atm; no gas circulation. (From Shihabi, 1981.)

mode would make an on-site application feasible without the need to build a hydrogen supply system. However, the concept requires extensive developmental study because there is scarcely any practical experience in on-site catalytic processing.

II. UPGRADING SHALE OILS

Synthetic crudes derived from shale oils typically possess pour points on the order of 25 to 35°C, too high to be marketed as distillate fuels because of their poor low temperature fluidity characteristics (Walsh, 1984; Angevine et al., 1984; LaPierre et al., 1986).

Table 8.1 Physical Properties of Shale Oils

	Analyses, wt %	
	In situ (Occidental)	Retort (Paraho)
Hydrogen	11.92	11.24
Nitrogen	1.45	1.86
Sulfur	0.60	0.71
Oxygen	0.90	1.30
Nickel	0.0012	0.00016
Iron	0.0065	0.0095
Arsenic	0.0028	0.0033
Bromine number	26.1	43.9
API gravity	23.4	20.5
Pour point, °C	16	27
Distillation (D1160), °C		
IBP	189	208
10%	253	245
50%	370	378
90%	512	474

Source: LaPierre et al. (1986).

Table 8.1 shows some typical properties of shale oils. It is noted that in addition to their high pour points, shale oils are also high in nitrogen and trace metals and generally require a hydrogenative pretreatment step before catalytic upgrading (Miller et al., 1982).

Catalysts prepared from an admixture of a medium pore zeolite and a porous refractory oxide supported with VIB and Group VIII metals have been designed to reduce pour point, nitrogen, and sulfur with minimum consumption of hydrogen (Ward and Carlson, 1986). Alternatively, the shale oil may be processed in a cascaded two-stage process (Gorring and Smith, 1979; LaPierre et al., 1986) that combines a first-stage hydrotreating reactor with a second-stage shape selective dewaxing reactor to yield a low pour point synthetic crude with minimal additional hydrogen consumption over the hydrotreating requirement. Results obtained with Paraho shale oil are shown in Figure 8.3. The product distribution was found to be dependent on the pour point of the liquid product. The yield of naphtha and LPG increased as the pour point of the liquid product decreased, the result of selective cracking of the paraffinic fraction of the feed.

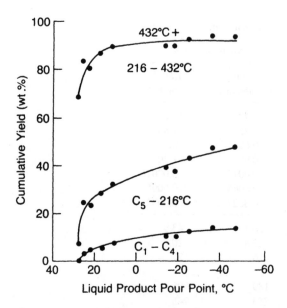

Figure 8.3   Hydrotreating/dewaxing yields—Paraho shale oil. (From LaPierre et al., 1986.)

The activity of the dewaxing catalyst at 425°C and above was found to be strongly dependent on the basic nitrogen content of the first-stage product and not affected by the high concentration of gaseous ammonia cascaded from the first-stage reactor to the dewaxing reactor, suggesting that the catalyst may function in this case as a nonacidic or low acidity catalyst.

In recent years the lack of commercial interest in upgrading shale oil has by and large eliminated new technological activity in this area.

III.   DEWAXING HYDROGEN DONOR SOLVENT FOR COAL LIQUEFACTION

Recent progress made by the use of hydrogen donor solvent for coal liquefaction through short contact time thermal reactions improves the selectivity in the conversion of coal to higher quality liquids and aids in the conservation of hydrogen, which is costly to produce (Whitehurst, 1980; Whitehurst et al., 1984). The effectiveness of a molecule as a hydrogen transfer agent depends strongly on the chemical structure of the molecule. Labile hydrogen of the hydroaromatics in the solvent are particularly effective in transferring hydrogen to stabilize the thermally cracked radical fragments. Thus, eight hydrogen atoms per molecule can be transferred from octahydrophenanthrene, while octadecane, a saturated hydrocarbon, has essentially no transferable hydrogen (Aiura et al., 1984).

In order to maintain the effectiveness of a hydrogen donor solvent, the buildup of *n*-paraffins in the recycle solvent in some of the coal liquefaction processes must be selectively removed (Yao et al., 1986a,b). A process known as the Nedol process, incorporating such concepts, is being developed in Australia and Japan under the Japanese New Energy Development Organization (NEDO) (Pace Synthetic Fuels Report, 1987) for the liquefaction of Victoria Brown coal and bituminous coal. Among the various processing schemes is the use of ZSM-5 for the dewaxing of the recycle solvent (Mitarai, 1987).

IV. CONVERTING NATURAL GAS TO LIQUID HYDROCARBONS

Natural gas, with over 3 quadrillion (10^{15}) cubic feet in estimated proven reserves worldwide (McCaslin, 1986), is another source of energy and chemical feedstock. At present, only a tiny fraction of the natural gas consumed is upgraded to more valuable products such as methanol and ammonia. Converting natural gas liquid transportation fuels could certainly become an important industry in the future as the supply of crude oil dwindles and the price differential between natural gas and the liquid fuels justifies it.

The synthesis of methanol from natural gas coupled with Mobil's Methanol-to-Gasoline (MTG) process is one of the newest technologies for the conversion of methane to liquid hydrocarbons using shape selective zeolites as the catalyst.

Another approach involves the direct conversion of natural gas to hydrocarbons by an oxidative coupling reaction (Sofranko et al., 1987). In this approach, methane is converted to higher molecular weight hydrocarbons by contacting it with a metal oxide in a cyclic process. In the methane reactor, the metal oxide is reduced while methane is converted to C_2^+ hydrocarbons and such byproducts as water, carbon oxides, and coke. The reduced oxide is then reoxidized in a separate reactor and circulated back to the methane reactor. The C_2^+ product can be upgraded to gasoline and distillates by coupling the methane reactor with Mobil's MOGD process (see Chapter 7, Section II) (Haxbun, 1985; Jones et al., 1986), again using shape selective zeolites as the

Table 8.2   Direct Partial Oxidation of CH_4 with O_2 over ZnZSM-5 at 65 atm, 7 vol % O_2, and 4600 h^{-1} GHSV

Reaction temperature, °C	460
CH_4 conversion, wt %	5.3
Carbon selectivity, wt %	
CO	71.7
CO_2	16.2
CH_3OH	2.5
Other oxygenates	0.2
C_2–C_4	1.7
C_5^+	7.7

Source: Han et al. (1994a).

catalyst. The critical step in these approaches is the activation of methane, whether by high temperature steam or by oxygen.

More recently, a thorough study of methane activation by direct partial oxidation (DPO) of methane over ZSM-5 and ZnZSM-5 catalysts was undertaken by Han et al. (1994a,b) with the objective to produce more C_5^+ hydrocarbons from CH_4 and O_2, but little success was achieved in obtaining useful selectivity (Table 8.2).

Wang et al. (1993) studied the catalytic performance of methane dehydrogenation and aromatization without using oxygen over MoHZSM-5. Their best result gave 5.6 wt % methane conversion at 700°C with an aromatics selectivity of more than 90% using a 2% Mo loaded HZSM-5 (Table 8.3); the conversion increased to 9% at 740°C (Xu et al., 1995).

Table 8.3   Conversion of Methane over MoHZSM-5 at HSV = 1500 h^{-1} g^{-1}

Reaction temperature, °C	700
CH_4 conversion wt %	5.6
Product selectivity[a], wt %	
Ethene	3.0
Ethane	2.2
Propane	0.1
Benzene	90.4
Toluene	4.4

[a]Yield of hydrogen not reported.
Source: Xu et al. (1995).

Other approaches to the activation process have also been studied. For example, Anderson and Tsai (1985) studied the activation of methane over HZSM-5 using nitrous oxide as the oxidant. As discussed in Chapter 4, methane can also be activated by reacting with halogen and sulfur. The resulting alkyl halides and mercaptans can be converted to higher molecular weight hydrocarbons over the medium pore molecular sieve catalysts (Chang et al., 1975a; Butter et al., 1975; Kaiser, 1987; Brophy et al., 1987). The difficulty with these latter approaches lies in the costly step of recovering the oxidants.

V. PRE-ENGINE CONVERTER

From the point of view of conserving global energy consumption, efforts to improve engine performance by using higher octane fuels could be counterbalanced by the loss of yield at the refinery in producing such fuels. CONCAWE (1983) published a study stating that the optimal unleaded gasoline octane at the refinery would be 94.5 R+0. While higher octanes can provide better engine performance, with current refining technology the demand of higher pool octane would require heavy capital investment at the refineries and decrease the gasoline yield per barrel of crude.

An alternative approach to this problem is to examine new technologies that could improve engine performance without the need of high octane fuels. Implementation of the pre-engine converter concept (Chen and Lucki, 1974) of attaching a catalytic reactor to an internal combustion engine and converting a low octane liquid fuel to a high octane gas/liquid fuel, eliminates the problems associated with producing high performance fuels at the refinery and shifts the task of reducing overall energy consumption back to the engine builders.

The concept of attaching a catalytic reactor to an internal combustion engine is not new. Cook (1940) disclosed a method of converting liquid hydrocarbons to gaseous fuel on board an engine. On-board production of hydrogen by steam reforming of gasoline fuel was studied by Newkirk and Abel (1972). However, the practicality of this concept using conventional catalytic systems was constrained in the past by the available space in a vehicle and the longevity of the catalyst. Medium pore zeolites, on the other hand, with their high activity and excellent stability have been demonstrated to fulfill these requirements (Chen and Lucki, 1974; Weisz et al., 1974; Chen, 1978; Degnan and Dessau, 1989; Chen and Degnan, 1989).

Shown in Figure 8.4 are the data on the octane boosting capability of ZSM-5 at 482°C as a function of space velocity. Figure 8.5 shows the stability

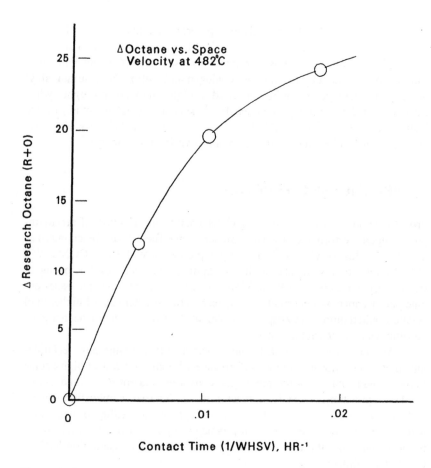

Figure 8.4 Δ octane versus space velocity at 482°C.

of the catalyst. The catalyst retains its octane boosting capability for over five hours of on-and-off operation.

The practical implementation of such a system, however, still awaits the development of advanced engine technology to satisfy such requirements as cold start, catalyst regeneration, and derivability. Recent advances in engine technology, such as computer control of air/fuel ratio, spark timing (Ikeura et al., 1980), and knock sensors (Currie et al., 1979) have gone a long way toward making this concept practical. However, there does not appear to be a great incentive for either the petroleum industry or the automobile industry to make a

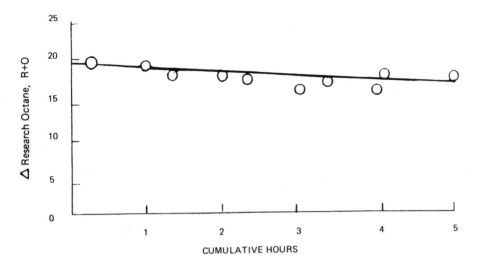

Figure 8.5 Δ octane versus on-stream time at 482°C, 100 WHSV.

move in that direction. Adoption of such an approach would drastically change the process of petroleum refining. From the point of view of conserving global energy consumption, technologies that could improve engine performance without the need of high-cotane fuels would eliminate the yield loss and the huge capital investment associated with producing high-performance fuels at the refineries. Adapting such a converter for methanol would also eliminate the need for the engine builders to produce methanol cars.

VI. LIQUID FUEL FROM BIOMASS

A. Direct Conversion of Carbohydrates

A number of technologies for the conversion of biomass to liquid fuels are currently available. They include, among others, fermentation, pyrolysis, hydrogenative liquefaction, and two-stage conversion processes such as gasification followed by alcohol synthesis or Fischer-Tropsch synthesis. Unfortunately, they either require large capital investment or are high in energy consumption and low in net carbon recovery (Weisz, 1982; Weisz and Marshall, 1984; Chen et al., 1986).

A significant improvement in the energy efficiency of any two-stage conversion process is possible if the two separate stages can be combined into a single stage such that the endothermic gasification reaction is coupled with the exothermic CO reduction reaction. In this single stage process, the removal of oxygen from the biomass occurs simultaneously with the enrichment of the hydrogen content of the liquid product. Furthermore, an improvement in the recovery of carbon is possible if the overall carbon-to-hydrogen ratio of the products can be reduced to below 2 (Chen et al., 1986).

An embryo of such a single-stage process is indicated by the catalytic conversion of sugars and hydrolyzed starch over HZSM-5. As was discussed in Chapter 4, when aqueous solutions of sugars and hydrolyzed starch were passed over HZSM-5 at 510°C, both hydrocarbons and CO were produced directly from these materials, combining the endothermic and exothermic reactions of a two-stage process. Examination of the data in Table 4.28 and Figure 4.24 shows that in addition to the hydrocarbon products, large quantities of CO were produced, which is convertible to hydrocarbons through Fischer-Tropsch synthesis or methanol synthesis. By counting the CO with the hydrocarbons, nearly 60% of the carbon in xylose is recovered as premium products, exceeding the net yield of current technologies.

Experiments with mixtures of carbohydrates and methanol showed that a synergism exists between carbohydrate and methanol, increasing the yield of hydrocarbons from the sugars (Figure 8.6). The best yield was with glucose, which produced 48.7 wt % carbon as hydrocarbons, 13.7 wt % as CO, and 33.4 wt % as coke, giving an estimated overall carbon recovery of 72.5 wt %, which is significantly higher than current technologies.

Unlike the current technology, which requires multi-step processing, the direct conversion process is nearly isothermal and therefore requires little process energy. But, of course, this is only the beginning; the process still has a long way to go before commercialization.

B. Biomass Pyrolysis

Pyrolysis of wood has been practiced for centuries to produce charcoal, wood alcohol, and other gas and liquid products. In more recent years, considerable international research efforts have been devoted to the development of methods to increase the yield of pyrolysis liquid (Antal, 1984, 1985) and to apply catalysis to upgrade the pyrolysis liquid to gasolinelike liquids (Chantal et al., 1984).

A number of new approaches to pyrolysis, including the use of short contact times (Scott et al., 1987; Diebold and Scahill, 1985, 1987a) and subatmospheric pressures (Lemieux et al., 1987), are being investigated. Primary pyrolysis product yields as high as 75 wt % of the dry wood converted have been reported (Diebold et al., 1986). These so-called primary pyrolysis products appear to be depolymerized fragments of the cellulose, hemicellulose, and lignin

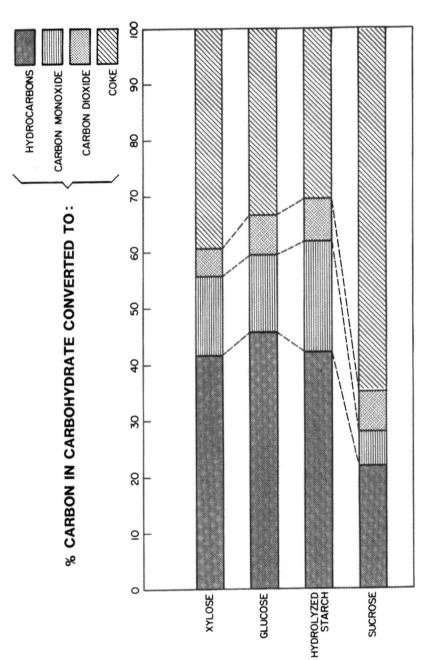

Figure 8.6 Production distribution from 4:1 methanol-carbohydrate mixtures at 510°C, 2 WHSV, 1 atm. (From Chen et al., 1986.)

present in the original wood (Evans et al., 1984). Their ultimate analysis also resembles that of the original wood, which has an effective hydrogen index near zero. However, they are highly reactive, forming tarry products upon standing, if not further processed. The upgrading of these unstable products to stable and more valuable fuels and chemicals represents a new challenge to catalysis.

Attempts to convert pyrolysis liquids to gasolinelike hydrocarbons over ZSM-5 (Chantal et al., 1984, Dao et al., 1987; Diebold and Scahill, 1987b; Chen et al., 1987) have given disappointingly low yields to be of practical interest. This is to be expected by their low effective hydrogen index. Higher yields can be expected only if some of the oxygen is retained in the final product or through selective decarbonylation and decarboxylation reactions to enrich the hydrogen content of the liquid product.

As discussed in Chapter 4, the conversion of other natural products that have higher effective hydrogen indices than carbohydrates, such as corn oil, peanut oil, castor oil and jojoba oil, over ZSM-5 has been reported by Weisz et al. (1979).

VII. APPLICATIONS IN OTHER INDUSTRIES

A. Fermentation

A process similar to the MTG process converting ethanol to hydrocarbons may be integrated into the ethanol fermentation process to produce fuels and chemicals from biomass. The integration can bring significant savings in energy cost because the energy-intensive rectifying columns may be eliminated by processing the overhead product from the beer column directly over the catalyst (Chen, 1983; Whitcraft et al., 1983; Aldridge et al., 1984; Anunziata et al., 1985).

Another possibility is to take advantage of the selective sorption properties of ZSM-5, which may be used to extract ethanol from the fermenter beer (Chen and Miale, 1983; Dessau, 1983; Dessau and Haag, 1984; Bul et al., 1985; Chen and Miale, 1985, 1987). By lowering the ethanol concentration in the fermenter with zeolites, it is conceivable that the rate of fermentation and the throughput of a fermenter can be significantly increased.

Sano et al. (1994b, 1995) proposed the use of a "silicalite" (a high silica-to-alumina ratio ZSM-5) membrane for the separation of alcohol/water mixtures and acetic/acid mixtures by molecular permeation.

B. Zeolitic Membrane Reactors

The development of a cohesive membrane of sufficient mechanical strength composed of a thin continuous layer of crystalline zeolite is not an easy task. A number of methods have been reported. For example, Haag and Tsikoyiannis (1991), Geus et al. (1993), Matsukata et al. (1994), and Sano et al. (1994a,b) crystallized a hydrogel for the synthesis of ZSM-5 on a variety of solid surfaces including alumina, metal surfaces, and polymeric surfaces. Yan et al. (1995) reported that ZSM-5 membranes with good permeation selectivity for *n*-butane-isobutane were grown from a clear solution on horizontally-held porous α-alumina disks.

The zeolitic membrane may also be used as a catalyst by first activating the membrane and then passing a feedstock through the wall made of the membrane when one of the reaction products has a higher permeability than the reactants. For example, when used in the dealkylation ethylbenzene, the high-permeation ethylene leads to a higher degree of dealkylation in greater selectivity.

C. Chemicals

In addition to the already commercialized processes described in Chapter 6, a number of other commercially useful chemicals can be produced with the shape selective zeolites. By taking advantage of the selectivity and stability of the catalyst, improved yield and process economics over the conventional technology can be expected.

Among various hydrocarbon products, medium pore molecular sieves should find application in the production of durene (Chang et al., 1975b), cumene (Kaeding, 1983), and *para*-dialkylbenzenes such as *p*-diethylbenzene (Ishida and Nakajima, 1986) and *p*-diisopropylbenzene (Kaeding, 1985). The selectivity and stability advantages may also be realized in the production of heterocompounds, such as alcohols (Chang and Morgan, 1980), glycols (Chang and Hellring, 1986), methyl *tert*-butyl ether (Chu and Kuehl, 1987), and the synthesis of selected isomers of aromatic amines (Chang and Lang, 1983, 1984), pyridines (Chang and Lang, 1980), alkylpyridines (Chang and Perkins, 1983), and alkylphenols (Kaeding et al., 1980; Young and Burress, 1980; Wu, 1983).

The isomerization activity of the medium pore zeolites for a large variety of aromatic and heterocyclic compounds should find applications in the production of difficult-to-synthesize isomers. For example, the *m*-cresol could be easily produced from its ortho and/or para isomers by isomerizing the latter isomers over the medium pore zeolites.

A novel route for the manufacture of *m*-cresol and 1,3-hydroxybenzene was proposed by Kaeding et al. (1980) starting with alkylation of toluene with propene to a mixture of meta/para isomers, followed by the selective dealkylation of the para isomer. The *m*-cymene is then converted to the final products by oxidation and rearrangement reactions, analogous to conditions for the commercial cumene-to-phenol process. The process avoids the cumbersome method of separating the meta isomer from an equilibrium cresol mixture.

As discussed in Chapter 4, the recent improvement in selectivity and stability of the synthesis of ϵ-caprolactam over the ZSM-5 catalyst by the Beckmann rearrangement of cyclohexanone oxime has the potential to replace the homogeneous process. ϵ-Caprolactam is one of the most important fiber precursors.

D. Environmental Applications

Polymer Waste Recovery

A new polymer waste recovery technology developed by Fuji Tech and Mobil has received some attention recently. The process combines catalytic conversion over ZSM-5 with thermal depolymerization of polymer wastes. Its commercialization should open a new front for the application of shape selective catalysis.

NO_x Abatement

Use of zeolites such as ZSM-5 and zeolite beta (Krishnamurthy et al., 1988) in the selective catalytic reduction (SCR) of NO and NO_2 can compete with nonzeolite catalysts in cleaning up combustion emissions from stationary sources. The zeolite-based catalysts are more efficient and less sensitive to contaminants than competing catalysts. They find applications in gas and diesel engines in power plants. In this process, ammonia is used as a reducing agent to convert NO_x to N_2 and H_2O. The presence of oxygen is essential since the reactions are believed to proceed according to the following reaction schemes:

$$2NO_2 + 4NH_3 + O_2 = 3N_2 + 6H_2O$$
$$4NO + 4NH_3 + O_2 = 4N_2 + 6H_2O$$

More recently, an intensive investigation has been undertaken, driven by the promise of controlling nitrogen oxides pollution from moving vehicles. ZSM-5 ion exchanged with metal ions such as Cu^{2+} was found to be much more active than without the Cu^{2+} exchange. Unfortunately, the presence of water as shown in Figure 8.7 strongly inhibits the reaction between NO and O_2 and SCR reactions (Iwamoto and Hamada, 1991; Shelef et al., 1994). Petunchi and Hall (1994) studied the possible causes of the irreversible damage of the catalyst. Since water is produced by the reaction, steaming can effect dealumination of the zeolite lattice, particularly at temperatures above 350°C. Armor (1994) commented that it is apparent that Cu-ZSM-5 is not the solution and more in-

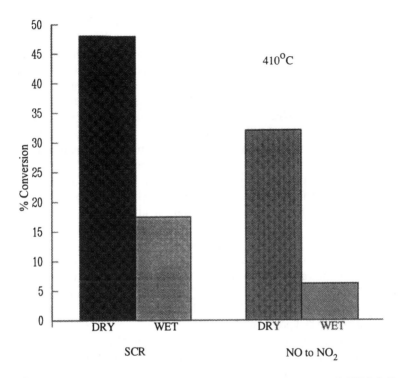

Figure 8.7 Effect of water on SCR and NO + O₂ reactions over CuZSM-5. Experimental conditions:

	SCR	NO + O₂
Catalyst, g	1.5	1.5
Inlet: NO_x, ppm	500	900
C_3H_6, ppm	1500	
O_2, %	4.9	4.9
Flow rate, liters/min	8.0	8.0

(From Iwamoto and Hamada, 1991; Shelef et al., 1993).

vestigators need to focus on new catalyst composition, not just modifying the particular composition.

E.   In Vivo Applications

The in vitro properties of ZSM-5 suggest its usefulness for in vivo applications. ZSM-5 is hydrophobic, acid stable (Garwood et al., 1988), nonfibrous

(Chu and Dwyer, 1983), and has a high selectivity for adsorbing ammonia and low molecular amines such as dimethylamine (Garwood and Chu, 1992), while excluding higher molecular weight compounds too large to fit in the pores. In the context used here, shape selective catalysis at the reaction temperatures relates to shape selective adsorption at body temperatures. Adding NaZSM-5 to the diet of Sprague-Dawley rats confirms that ZSM-5 binds ammonia ($NH_3 + NH_4^+$) with no observable harmful effects as noted by Visek and Mangian at the College of Medicine, University of Illinois (Garwood et al., 1994). Ammonia has been demonstrated to promote colon carcinogenesis in rats (Clinton et al., 1988). The high selectivity of ZSM-5 for adsorbing dimethylamine might also be useful in preventing the formation of N-nitrosodimethylamine, a known carcinogen formed in the gastrointestinal tract by reaction with nitrous acid (Mergens and New-mark, 1981).

Natural zeolites, primarily clinoptilolite, has been studied for decades as an additive to the feed of animals, with beneficial effects on health (Mumpton, 1984). However, the varying composition of samples taken from the earth has discouraged clinical application (Pond and Yen, 1984). The discouragement should be overcome by using ZSM-5, which is a synthetic material of repro-ducible composition and properties.

Described here is the study of the effect of adding NaZSM-5 to the diet of Sprague-Dawley rats conducted by Viek and Mangian (Garwood et al., 1994). The testing procedure is outlined in Table 8.4.

Table 8.4   Procedure of Testing with Rats

- Male Sprague-Dauley rats weaned, initial weight: 40 g.
- Individually caged in stainless steel cages at $22 \pm 1°C$ with 14 h fluorescent light per 24 h.
- Basal diet fed ad libitum to all animals for 4 days.
- Randomly assigned to 5 treatment groups (8 per group), 0.625, 1.25, 2.5, and 5.0 wt % NaZSM-5 added to Basal diet.
- Fresh diet provided twice weekly with ad libitum feeding and free access to distilled water for 28 days.
- On 21st day, urine and feces quantitatively collected separately for 3 days, analyzed for nitrogen.
- On 29th day, rats euthanized, small and large intestines excised, contents immediately released under 6N H_2SO_4, homogenized, centrifuged, supernatant liquid analyzed for ammonia (Chaney and Marchant, 1962).

Source: Garwood et al. (1994).

An important result is that the zeolite is binding nitrogen, as shown by the nitrogen concentration in feces increasing with the amount of NaZSM-5 in the feed (Figure 8.8).

The second significant result is the increase of the amount of ammonia nitrogen ($NH_3 + NH_4^+$) in the intestines with the weight percent of NaZSM-5 in the feed, as shown in Figure 8.9. These results confirm conclusively the ammonia is indeed bound by the NaZSM-5 in vivo in the gastrointestinal tract of living animals.

Relative to the dosage and the cost of other feed additives, such as antibiotics, the current cost of NaZSM-5 synthesis is too high. However, considering the annual feed production for animals in the United States is so high (Table 8.5), adding 0.5 wt % of NaZSM-5, or 10 lb/ton, to feed will call for the production of 5.5 million tons of ZSM-5 per year—a volume much larger than is currently used for industrial catalyst—which should give sufficient incentive to reduce the cost of synthesis.

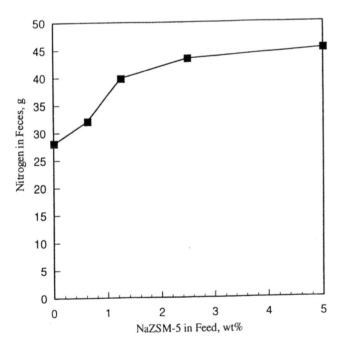

Figure 8.8 Effect of NaZSM-5 in feed on fecal nitrogen of Sprague-Dawley rats. (From Garwood et al., 1994.)

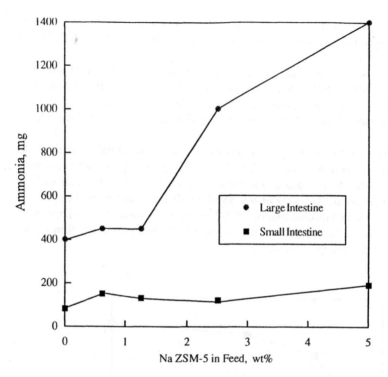

Figure 8.9 Effect of NaZSM-5 in feed on ammonia nitrogen ($NH_3 + NH_4^+$) in intestines of Sprague-Dawley rats. (From Garwood et al., 1994.)

Table 8.5   Annual feed production for animals in U.S.

	Tons (millions)
Poultry, including turkey	50
Swine	15
Dairy cattle	18
Beef cattle	18
Others	9
Total	110

Source: W. J. Visek (personal communication, July 1993).

REFERENCES

Aiura, M., T. Masunaga, K. Moriya, and Y. Kageyama, Fuel *63*, 1138 (1984).

Aldridge, G. A., X. E. Veryklos, and R. Mutharasan, Ind. Eng. Chem. Process Des. Dev. *23*, 733, (1984).

Anderson, J. R. and P. Tsai, Appl. Catal. *19*, 141 (1985).

Angevine, P. J., G. H. Kuehl, and S. Mizrahi, U.S. Pat. 4,431,518, Feb. 14, 1984.

Anunziata, O. A., O. A. Orio, E. R. Herrero, A. F. Lopez, and A. R. Suarez, Appl. Catal. *15*, 235 (1985).

Antal, M. J., Jr., "Biomass Pyrolysis. A Review of the Literature—Part 1: Carbohydrate Pyrolysis," in *Advan. Solar Energy*, *1*, 61 (1982), K. W. Boer and J. A. Duffie, eds., Am. Solar Energy Soc., Boulder, CO.

Antal, M. J., Jr., "Biomass Pyrolysis. A Review of the Literature—Part 2: Lignocellulose Pyrolysis," in *Advan. Solar Energy*, *2*, 175 (1985), K. W. Boer and J. A. Duffie, eds., Am. Solar Energy Soc./Plenum Press, New York.

Armor, J. N., Appl. Catal. *B4*, N18 (1994).

Brophy, J. H., J. J. Fontfreide, and J. D. Tomkinson, U.S. Pat 4,652,688, Mar. 24, 1987.

Bul, S., X. Veryklos, and R. Mutharasan, Ind. Eng. Chem. Process Des. Dev. *24*, 1209 (1985).

Butter, S. A., A. T. Jurewicz, and W. W. Kaeding, U.S. Pat. 3,894,107, Jul. 8, 1975.

Chang, C. D. and S. D. Hellring, U.S. Pat. 4,620,044, Oct. 28, 1986.

Chang, C. D. and W. H. Lang, U.S. Pat. 4,220,783, Sep. 2, 1980.

Chang, C. D. and W. H. Lang, U.S. Pat. 4,380,669, Apr. 19, 1983.

Chang, C. D. and W. H. Lang, U.S. Pat. 4,434,299, Feb. 28, 1984.

Chang, C. D. and N. J. Morgan, U.S. Pat. 4,214,107, Jul. 22, 1980.

Chang, C. D., W. H. Lang, and A. J. Silvestri, U.S. Pat. 3,894,104, Aug. 9, 1975a.

Chang, C. D., A. J. Silvestri, and R. L. Smith, U.S. Pat. 3,894,105, Aug. 9, 1975b.

Chang, C. D. and P. D. Perkins, U.S. Pat. 4,388,461, Jun. 14, 1983; U.S. Pat. 4,395,554, Jul. 26, 1983.

Chantal, P., S. Kaliaguine, J. L. Grandmaison, and A. Mahey, Appl. Catal. *10*, 317 (1984).

Chen, N. Y., U.S. Pat. 4,070,993, Jan. 31, 1978.

Chen, N. Y., Chemtech *13*, 488 (1983).

Chen, N. Y. and T. F. Degnan, U.S. Pat. 4,862,836, Sep. 5, 1989.

Chen, N. Y. and S. J. Lucki, "Approaches to Automotive Emissions Control," R. W. Hurn, ed., Am. Chem. Soc. Symp. Ser. *1*, 69 (1974).

Chen, N. Y. and J. N. Miale, U.S. Pat. 4,420,561, Dec. 13, 1983.

Chen, N. Y. and J. N. Miale, U.S. Pat. 4,515,892, May 7, 1985.

Chen, N. Y. and J. N. Miale, U.S. Pat. 4,690,903, Sep. 1, 1987.

Chen, N. Y. and D. S. Shihabi, U.S. Pat. 4,269,697, May 26, 1981.

Chen, N. Y., T. F. Degnan, Jr., and L. R. Koenig, Chemtech *16*, 506 (1986).

Chen, N. Y., D. E. Walsh, and L. R. Koenig, "Fluidized Bed upgrading of Wood Pyrolysis Liquids and Related Compounds," Paper 67, Symposium on Cellulose and

Related Materials, American Chemical Society 193rd National Meeting Denver, Apr. 5–10, 1987.

Chu, P. C. and G. H. Kuehl, Ind. Eng. Chem. Res. *26*, 365 (1987).

Chu, P. and F. G. Dwyer, "Inorganic Cation Exchange Properties of Zeolite," in *Intrazeolite Chemistry*, G. D. Stucky and F. G. Dwyer, eds., Am. Chem. Soc. Symp. Ser. *218*, 59 (1983).

Clinton, S. K., D. G. Bostwick, L. M. Olson, H. J. Mangian, and W. J. Visek, Cancer Research *48*, 3035 (1988).

CONCAWE Report, "Assessment of the Energy Balances and Economic Consequences of the Reduction and Elimination of Lead in Gasoline," CONCAWE. The Hague, Dec., 1983.

Cook, J. T., U.S. Pat. 2,201,965, May 21, 1940.

Currie, J. H., D. S. Grossman, and J. J. Gumbleton, "Energy Conservation with Increased Compression Ratio and Electronic Knock Control," Paper 790173, SAE Meeting, Detroit, Feb. 26–Mar. 2, 1979.

Dao, L. H., M. Haniff, A. Houle, and D. Lamothe, "Reactions of Biomass Pyrolysis Oils Over ZSM-5 Zeolite Catalysts," Paper 71, Symposium on Cellulose and Related Materials, Am. Chem. Soc. 193rd Nat. Mtg., Denver, Apr. 5–10, 1987.

Degnan, T. F. and R. M. Dessau, U.S. Pat. 4,884,531, Dec. 5, 1989.

Dessau, R. M., U.S. Pat. 4,423,280, Dec. 27, 1983.

Dessau, R. M. and W. O. Haag, U.S. Pat. 4,442,210, Apr. 8, 1984.

Diebold, J. P. and J. W. Scahill, Entrained-Flow Fast Ablative Pyrolysis of Biomass," Annual Report, Oct. 1, 1983–Nov. 30, 1984, Solar Res. Inst., Golden, CO, SERI/PR-2665, 1985.

Diebold, J. P. and J. W. Scahill, "Production of Primary Pyrolysis Oils in a Vortex Reactor," Paper 10, Symposium on Cellulose and Related Materials, Am. Chem. Soc. 193rd Nat. Mtg. Denver, Apr. 5–10, 1987a.

Diebold, J. P. and J. W. Scahill, "Biomass to Gasoline (BTG): Upgrading Pyrolysis Vapors to Aromatic Gasoline with Zeolite Catalysts at Atmospheric Pressure," Paper 70, Symposium on Cellulose and Related Materials, Am. Chem. Soc. 193rd Nat. Mtg. Denver, Apr. 5–10, 1987b.

Diebold, J. P., H. L. Chum, R. J. Evans, T. A. Milne, T. B. Reed, and J. W. Scahill, "Low-Pressure Upgrading of Primary Pyrolysis Oils from Biomass and Organic Wastes," paper presented at the 10th IGT/CBETS Conference on Energy from Biomass and Wastes, Washington, April 7–10, 1986.

Evans, R. J., T. A. Milne, and M. N. Soltys, J. Anal. Appl. Pyrol. *6*, 273 (1984).

Garwood, W. E. and P. Chu, Catalysis of Organic Reactions, *47*, 365 (1992).

Garwood, W. E., P. Chu, W. J. Visek, and H. J. Mangian, "Relationship of in Vitro Properties of Medium Pore Zeolite ZSM-5 to in Vivo Performance," Paper 326, Inorganic Div., Am. Chem. Soc. 207th Nat. Mtg., San Diego, Mar. 12–17, 1994.

Garwood, W. E., P. Chu, N. Y. Chen, and J. C. Bailar Jr., Inorg. Chem. *27*, 9331 (1988).

Geus, E. R., H. van Bekkum, J. W. Bakker, and J. A. Moulijn, Microporous Mater. *1*, 131 (1993).

Gorring, R. L. and R. L. Smith, U.S. Pat. 4,153,540, May 8, 1979.

Haag, W. O. and J. G. Tsikoyiannis, U.S. Pat. 5,069,704, Dec. 3, 1991.

Han, S., D. J. Martenak, R. E. Palermo, J. Pearson, and D. E. Walsh, J. Catal. *148*, 134 (1994a).

Han, S., E. A. Kaufman, D. J. Martnek, R. E. Palermo, J. Pearson, and D. E. Walsh, Catal. Lett. *29*, 27 (1994b).

Haxbun, E. A., U.S. Pat. 4,556,749, Mar. 12, 1985.

Ikeura, K., A. Hosaka, and T. Yano, "Microprocessor Control Brings About Better Fuel Economy with Good Driveability," Paper 800056, SAE Meeting, Detroit, Feb. 25–29, 1980.

Ishida, H. and H. Nakajima, U.S. Pat. 4,613,717, Sep. 23, 1986.

Iwamoto, M. and H. Hamada, Catal. Today *10*, 57 (1991).

Jones, C. A., J. J. Leonard, and J. A. Sofranko, U.S. Pat. 4,567,307, Jan. 28, 1986.

Kaeding, W. W., U.S. Pat. 4,393,262, Jul. 12, 1983.

Kaeding, W. W., Eur. Pat. 148,584, Jul. 17, 1985; Eur. Pat. 149,508, Jul. 24, 1985.

Kaeding, W. W., M. M. Wu, L. B. Young, and G. T. Burress, U.S. Pat. 4,197,413, Apr. 8, 1980.

Kaiser, S. W., U.S. Pat. 4,677,243, Jun. 30, 1987.

Krishnamurthy, S., J. P. Williams, D. A. Pappal, C. T. Sigal, and T. R. Killiany, U.S. Pat. 4,778,665, Oct. 18, 1988.

Lemieux, R., C. Roy, B. de Caumia, and D. Blanchette, "Preliminary Engineering Data for Scale-Up of a Biomass Vacuum Pyrolysis Reactor," Paper 9, Symposium on Cellulose and Related Materials, Am. Chem. Soc. 193rd Nat. Mtg. Denver, Apr. 5–10, 1987.

LaPierre, R. B., R. L. Gorring, and R. L. Smith, "Extended End-Point Distillate Fuels from Shale Oil by Hydrotreating Coupled with Catalytic Dewaxing," paper presented at the Am. Chem. Soc. Nat. Mtg., New York, Apr. 17, 1986.

Matsukata, M., N. Nishiyama, and K. Ueyama, J. Chem. Soc., Chem. Commun. *3*, 227 (1994).

McCaslin, J. C., ed., *International Petroleum Encyclopedia*, Vol. 19, PennWell, Tulsa, Okla., p. 294 (1986).

Mergens, W. J. and H. C. Newmark, "Blocking Nitrosation Reactions In Vivo," in *N-Nitroso Compounds*, R. A. Scalan and S. R. Tannenbaum, eds., Am. Chem. Soc. Symp. Ser. *174*, 193 (1981).

Miller, W., M. Harvey and M. Hunter, "Premium Syncrude from Oil Shale Using Union Oil Technology," paper presented at the NPRA Annual Meeting, San Antonio, Mar. 1982.

Mitarai, Y., "Catalyst Research for Solvent Hydrotreating in Nedol Process," Paper B-51, 3rd China-Japan-USA Symposium on Catalysis, Xiamen, Aug. 7–11, 1987.

Mumpton, F. A., "The Role of Natural Zeolites in Agriculture and Aquaculture" in *Zeo-Agriculture*, W. G. Pond and F. A. Mumpton, eds., Boulder, CO., p. 3, 1984.

Newkirk, M. S. and J. L. Abel, U.S. Pat. 3,682,142, Aug. 8, 1972.

Pace Synthetic Fuels Report *24(1)*, 4-72, J. E. Sinor, ed., Pace Consultants, Houston (March 1987).

Petunchi, J. O. and W. K. Hall, Appl. Catal. *B3*, 239 (1994).

Pond, W. G., and J. T. Yen, "Physiological Effects of Clinoptilolite and Synthetic Zeo-lite A in Animals," in *Zeo-Agriculture*, W. G. Pond and F. A. Mumpton, eds., Boulder, CO., p. 128 (1984).

Sano, T., Y. Kiyozumi, K. Maeda, M. Toa, S. Niwa, and F. Mizukami, J. Mater. Chem. *2*, 141 (1994a).

Sano, T., M. Hasegawa, Y. Kawakami, Y. Kiyozumi, H. Yanagishita, D. Kitamoto, and F. Mizukami, Stud. Sur. Sci. Catal. *84*, 1175 (1994b).

Sano, T., S. Ejiri, M. Hasegawa, Y. Kawakami, N. Enomoto, Y. Tamai, and H. Yanag-ishita, Chem. Lett. *2*, 153 (1995).

Scott, D. S., J. Piskorz, A. Grinshpun, and R. G. Graham, "Effect of Temperature on Liq-uid Product Composition from the Fast Pyrolysis of Cellulose," Paper 7, Sym-posium on Cellulose and Related Materials, Am. Chem. Soc. 193rd Nat. Mtg. Denver, Apr. 5–10, 1987.

Shelef, M., C. N. Montreuil, and H. W. Jen, Catal. Lett. *26*, 277 (1994).

Shihabi, D. S., U.S. Pat. 4,284,529, Aug. 18, 1981.

Shihabi, D. S., U.S. Pat. 4,377,469, Mar. 3, 1983.

Sofranko, J. A., J. J. Leonard, C. A. Jones, A. M. Gaffney, and H. P. Withers, "Catalytic Oxidative Coupling of Methane Over Sodium-Promoted Manganese Oxide on Silica and Magnesia," Paper C-8, 10th N. Am. Mtg. Catal. Soc., San Diego, May 17–22, 1987.

Walsh, D. E., U.S. Pat. 4,428,826, Jan. 31, 1984.

Wang, L., L. Tao, M. Xie, G. Xu, J. Huang, and Y. Xu, Catal. Lett. *21*, 35 (1993).

Ward, J. W. and T. L. Carlson, U.S. Pat. 4,600,497, Jul. 15, 1986.

Weisz, P. B., N. Y. Chen, and S. J. Lucki, U.S. Pat. 3,855,980, Dec. 24, 1974.

Weisz, P. B., Chemtech *12*, 114 (1982).

Weisz, P. B. and J. F. Marshall, *Fuels from Biomass: A Critical Analysis of Technology and Economics*, Marcel Dekker, New York, 1984.

Weisz, P. B., W. O. Haag, and P. G. Rodewald, Science *206*, 57 (1979).

Whitcraft, D. R., X. E. Veryklos, and R. Mutharasan, Ind. Eng. Chem. Process Des. Dev. *22*, 452 (1983).

Whitehurst, D. D., Am. Chem. Soc. Symp. Ser. *139*, 133 (1980).

Whitehurst, D. D., T. O. Mitchell, and M. Farcasiu, "Coal Liquefaction," p. 274, Acad-emic Press, New York, 1984.

Wise, J. J., N. Y. Chen, and J. R. Katzer, paper presented at Am. Chem. Soc. Mtg., New York, Apr., 1986.

Wu, M. M., U.S. Pat. 4,391,998, Jul. 5, 1983.

Xu, Y., S. Liu, L. Wang, M. Xie, and X. Guo, Catal. Lett. *30*, 135 (1995).

Yan, Y., M. Tsapatsis, G. R. Gavalas and M. E. Davis, J. Chem. Soc. Chem. Commun. *2*, 227 (1995).

Yao, T., K. Hayakawa, K. Kurachi, and T. Chikada, Japan Pat. 61-181890, Aug. 14, 1986a.

Yao, T., K. Hayakawa, and K. Kurachi, Japan Pat. 61-203197, Sep. 9, 1986b.

Young, L. B. and G. T. Burress, U.S. Pat. 4,205,189, May 27, 1980.

9

Conclusions

Shape selective catalysis has made giant strides since its inception in the late 1950s. From the original discovery that selective conversion of linear molecules can be achieved by molecular size exclusion of nonlinear molecules from the catalytic sites, more subtle ways of achieving catalytic selectivity were discovered in the 1960s making use of such phenomena as configurational diffusional constraints and spatiospecific or transition state inhibition. With the discovery of new catalytic materials, such as the medium pore zeolites and a large variety of zeolites in the "dual pore system," catalysts have been designed that can discriminate not only linear molecules from branched molecules, but also molecules of varying degrees of chain branching and certain isomers of aromatic and naphthenic molecules, many of which are of industrial importance. Among the metallophosphate zeolites ($AlPO_4$, SAPO, TAPO, MeAPO, etc.) several have found industrial process applications.

As in the case of the 8-membered oxygen ring zeolites, the shape and size of the 10- or 12-membered oxygen rings also vary from one structural type to another. They range from nearly circular to elliptical to odd shapes such as tear drops. Industrial catalytic processing employing shape selective catalysts are now a reality in petroleum refining and petrochemical industries. New catalytic dewaxing processes have created such new products as extended boiling range premium jet fuels and diesel fuels, and ultra–low temperature paraffinic refrigeration oils and lubricating oils. A number of catalytic lube dewaxing processes are replacing the solvent dewaxing process for their high product quality, cost-effectiveness in energy saving, and operating simplicity. Operating in conjunction with catalytic re-forming and catalytic cracking processes, new methods using shape selective catalysts have broken the octane barriers of traditional processes for producing gasoline, thereby lowering benzene content in the gasoline to meet environmental standards and offering new alternative processing schemes to refine crudes that will make more efficient use of hydrogen and increase a refiner's ability to handle hydrogen-deficient heavy crudes.

The design of shape selective aromatics processing catalysts have created new processing routes to BTX from a variety of starting materials including light gases. Several alumina-based and aluminum chloride–based catalysts are being replaced by shape selective catalysts for their superior activity, stability, and product selectivity in the production of xylenes and ethyl benzene, major starting materials for the synthetic fiber and polymer industries.

Similar advances have been demonstrated and commercialized in the application of shape selective catalysis to the production of synthetic fuels and chemicals. Notable among them are the methanol-to-gasoline (MTG) process, the methanol-to-olefin (MTO) process, and the demonstrated ability to produce a large variety of hydrocarbon and nonhydrocarbon chemicals.

Since the publication of the original edition of this book, several zeolites have found commercial applications, including Linde Type L, ZSM-12, zeolite Beta, SAPO-11, and MCM-22. New materials such as EMC-2, SSZ-26, and SSZ-33 have been discovered. Many methods of synthesis and modification have produced useful zeolites, including boron-, gallium-, and titanium-modified zeolites, making their coverage a significant addition to this book. With the ever expanding knowledge of the synthesis and characterization of novel new materials, opportunities for new catalytic chemistry and new industrial applications can continue to be expected.

There is, however, an apparent interdisciplinary gap between the study of zeolite catalysis and industrial applications. Joint industrial and academic efforts would be most helpful in expanding knowledge in these areas, potentially leading to opportunities for new catalytic chemistry and new industrial applications. Finally, the establishment of an international materials research center to synthesize and characterize known zeolites would promote the use of consistent zeolite samples throughout the scientific community. The availability of these samples would generate a wealth of fundamental knowledge on zeolite properties, which would be a most desirable avenue for progress.

Author Index

Subject Index